PROPERTY OF
MOUNT ROYAL P.S.
LIBRARY

LIBRARY

DISCARDED

Survivor's Science in the Rain Forest

Peter D. Riley

Raintree
Chicago, Illinois

© 2004 Raintree
Published by Raintree, a division of Reed Elsevier, Inc.
Chicago, Illinois
Customer Service 888-363-4266
Visit our website at www.raintreelibrary.com

All rights reserved. No part of this book may be reproduced or transmitted in any form or by any means, electronic or mechanical, including photocopying, recording, taping, or any information storage and retrieval system, without permission in writing from the publisher.

For information, address the publisher:
Raintree, 100 N. LaSalle, Suite 1200, Chicago, IL 60602

Library of Congress Cataloging-in-Publication Data:

Riley, Peter D.
  Survivor's science in the rain forest / Peter D. Riley.
     v. cm. -- (Survivor's science)
Includes bibliographical references and index.
Contents: Rain forests of the world -- Why are rainforests hot and wet? -- Clothes for the rainforest -- Rainforest plants -- Traveling through the rainforest -- Finding a way -- Making a shelter -- Fire -- Water -- Food -- Dangers to health -- Rain forest animals -- Rescue.
  ISBN 1-4109-0227-7 (lib. bdg.-hardcover)
 1. Wilderness survival--Juvenile literature. 2. Rain forest ecology--Juvenile literature. [1. Wilderness survival. 2. Rain forests. 3. Rain forest ecology. 4. Ecology.] I. Title. II. Series.
  GV200.5.R55 2004
  613.6'9--dc21
                                   2003008624
Printed in Italy.
Bound in the United States.
07 06 05 04 03
10 9 8 7 6 5 4 3 2 1

Acknowledgments
The publishers would like to thank the following for permission to reproduce photographs and illustrations:
p. 4 Roland Seitre/Still Pictures; p. 7 (bottom) Kevin Schafer/Corbis; p. 9 Patrick Field/Eye Ubiquitous; p. 10 Jacques Jangoux/Still Pictures; p. 12 Wendy Stone/Corbis; p. 15 Dr. John Brackenbury/Science Photo Library; p. 16 Layne Kennedy/Corbis; p. 18 Imtiaz Hussein/Eye Ubiquitous; pp. 19, 20 Wolfgang Kaehler/Corbis; p. 22 Chris Caldicott/Still Pictures; p. 26 Peter Kerslake/Eye Ubiquitous; p. 28 Norbert Schaefer/Corbis; p. 31 Dave Bartruff/Corbis; p. 34 Gary Braasch/Corbis; p. 35 Dani/Jeske/Still Pictures; p. 36 Dario Novellino/Still Pictures; p. 37 (top) Dario Novellino/Still Pictures; p. 37 (bottom) Alison Wright/Corbis; p. 38 Daniel Heuclin/Still Pictures; p. 40 (left) Kevin Schafer/Corbis; p. 40 (right) Patricia Fogden/Corbis; p. 42 Alain Compost/Still Pictures; p. 44 Luiz C. Marigo/Still Pictures. The science activity photographs and all illustrations are by Carole Binding.

Cover Photographs: Dario Novellino/Still Pictures and Luiz C. Marigo/Still Pictures.

Every effort has been made to contact copyright holders of any material reproduced in this book. Any omissions will be rectified in subsequent printings if notice is given to the publisher.

Content Consultant:
Karen Carney is a graduate student in Learning Science at Northwestern University. Prior to this she was an Earth Science and Geology teacher at Collegiate School in New York City. Karen has worked as a geological field and lab researcher, specializing in fossil preservation and sedimentary environments.

Some words are shown in bold, **like this.** you can find out what they mean by looking in the glossary.

Contents

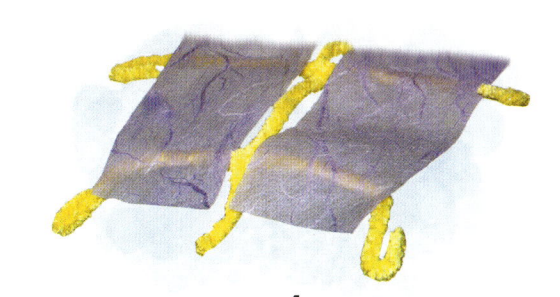

Introduction	4
Rain Forests of the World	6
Why are Rain Forests Hot and Wet?	8
Clothes for the Rain Forest	12
Which cloth dries fastest?	13
Rain Forest Plants	16
Do seeds grow better in rain forest conditions?	17
Traveling Through the Rain Forest	19
Make a model dugout canoe	21
Finding a Way	22
Make a compass	25
Making a Shelter	26
Tie a timber hitch knot	27
Extracting plant fibers	29
Building a Campfire	30
Water to Drink	32
Can you make a filter from a sock?	33
Food	36
Dangers to Health	38
Rain Forest Animals	40
Make a simple resonator	41
How do skin folds affect a flying squirrel?	43
Rescue	44
Testing aircraft wing shapes	45
Glossary	46
Index	48

Introduction

A huge number of species of animals (possibly a few million) live in the rain forests. Scientists go on expeditions to study them.

What's it like in a rain forest?

Imagine going into a bathroom after someone has had a hot shower. The air is very warm and damp. If you switched off the light and closed the curtains and door to make the bathroom gloomy, you would have a sense of what it is like on the rain forest floor during the day. At night it is cooler and totally dark. It is also noisy, as rain forest animals call to each other across the trees.

How would you cope in the rain forest? As you moved through this hot, wet **habitat**, you would quickly become hot and wet (from sweat) yourself. Although there is plenty of rain, much of the water is not safe to drink, and some of the fruits you might think of eating are poisonous. The rain forest can be dangerous if you are not prepared.

The keys to survival

Thousands of years ago some people moved into the rain forests and learned how to survive there. In more recent times, explorers have made **expeditions** into the rain forests to make maps and find out about the wildlife. Before an expedition, they plan how they will survive. They study how to make a shelter, how to find safe water and food, what to wear, how to travel safely, how to find the way, and, if there is an emergency, how to be rescued.

Discovering with science

For thousands of years people have investigated their surroundings and made discoveries that have helped them survive. About 400 years ago, a way of investigating called the scientific method was devised to help us understand our world more clearly. The main features of the scientific method are:

1 Making an **observation**

2 Thinking of an idea to explain the observation

3 Doing a test or experiment to test the idea

4 Looking at the result of the test and comparing it with the idea

Today the scientific method is used to provide explanations for almost everything. In this book you can find out about the science that helps people, plants, and animals survive in rain forests. You can also try the activities shown here to see how different aspects of science help life survive in the hottest and wettest places on Earth. In these activities you may use all the steps in the scientific method or just parts of it, such as making observations or doing experiments. But you will always be using science to make discoveries.

Are you ready to find out how people survive in the rain forest? Turn the page to find the major rain forests of the world.

Test clothes for the rain forest.
Page 13

Compare plant growth.
Page 17

Investigate buoyancy.
Page 21

Make a compass.
Page 25

Observe plant fibers.
Page 29

Investigate forces in knots.
Page 27

Test a filter.
Page 33

Investigate resonating sound.
Page 41

Test air resistance.
Page 43

Make an airfoil wing.
Page 45

Rain Forests of the World

The four main rain forest regions are in the tropics, a part of the earth around the equator. The weather here is hot, with very heavy rainfall.

Many types of plants and animals live in one rain forest region only and are not found anywhere else. Each rain forest is also the home of groups of people who have learned how to survive in this special environment. The rain forest provides them with food, drink, clothing, building materials, and medicines.

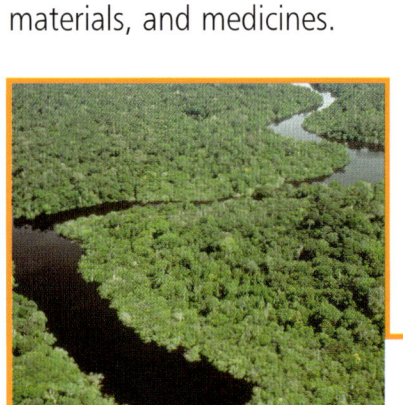

The tropics

Rain forest regions

Tropic of Cancer

Equator

Amazon River

South America

Tropic of Capricorn

The Amazon rain forest

This is the largest rain forest in the world. It covers a huge area (as big as the United States) around the Amazon River and its tributaries. Rain collects in the rivers and flows out of the Amazon into the ocean. After heavy rain, the water that flows out of the Amazon would fill 12,000,000 buckets every second.

Many plants and animals here are unusual. This is because South America was once an island, so its plants and animals developed in a different way from living things in other parts of the world. Howler monkeys and brightly colored parrots live in the treetops. Frogs and insects are found at every level, while jaguars and armadillos live on the forest floor.

There are several groups of people that live here, such as the Waiwai and Kayap'o.

The African rain forest

This rain forest grows around the Congo River and its many tributaries.

This is home for the gorilla, which feeds on plant material such as fruit, bark, and roots. The okapi is like a small, short-necked giraffe. It makes paths through the vegetation and feeds on a wide variety of plants.

The Pygmies are one type of people who live in the African rain forest.

Rain forests of Southeast Asia

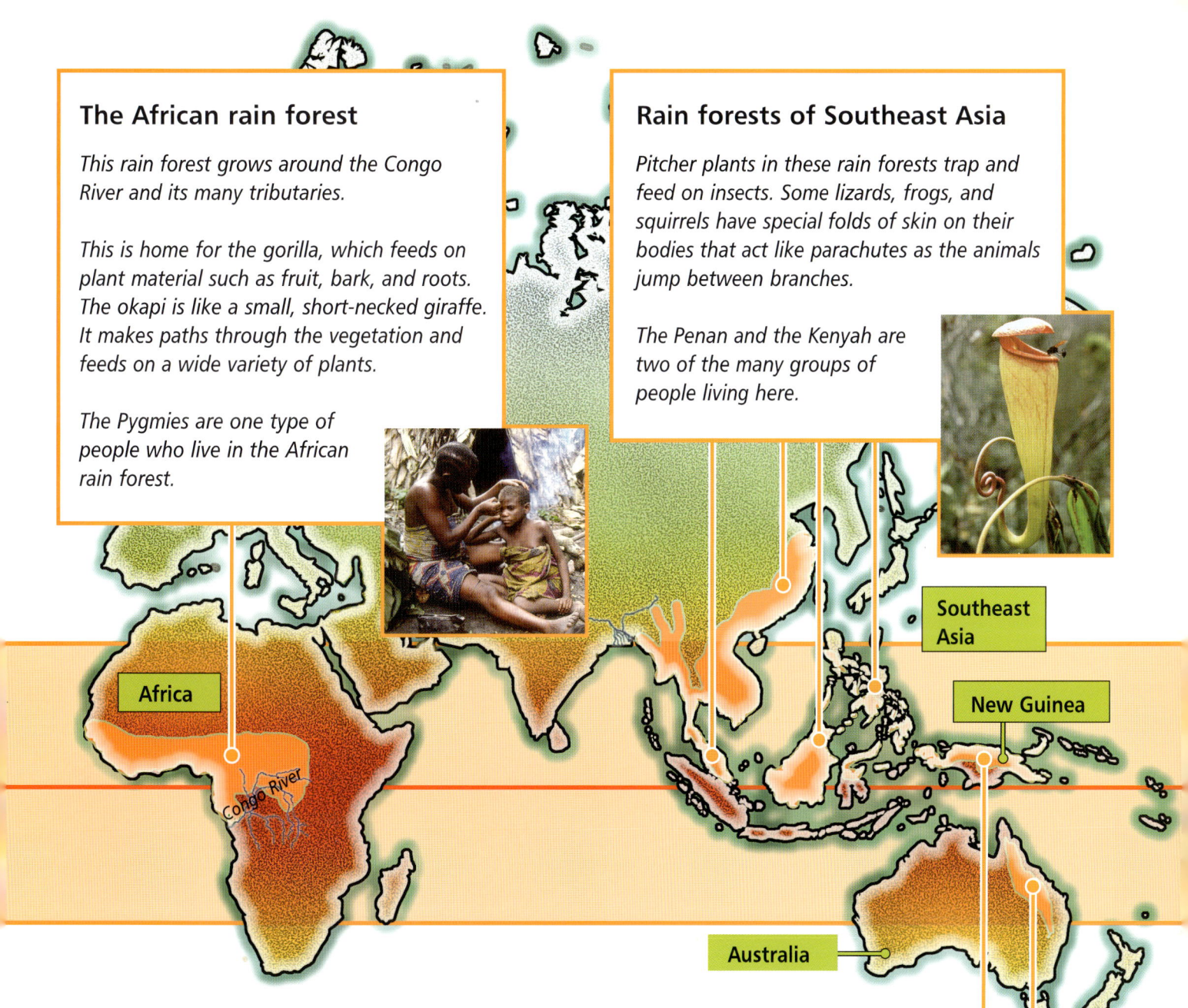

Pitcher plants in these rain forests trap and feed on insects. Some lizards, frogs, and squirrels have special folds of skin on their bodies that act like parachutes as the animals jump between branches.

The Penan and the Kenyah are two of the many groups of people living here.

The rain forests of New Guinea and Australia

Here you may see a tree kangaroo, or maybe a cassowary (left), a large flightless bird that lives on the ground. The cassowary could kill a person with one kick if attacked, but normally hides in the vegetation. It eats many kinds of fruit.

In the rain forests of New Guinea, the Huli and the Mendi people make wigs from human hair, feathers, and flowers to wear when they gather for special occasions. In the Australian rain forests, **Aborigine** groups like the Djabuganjdiji are known for their music, dances, and paintings.

Why are Rain Forests Hot and Wet?

Heat from the sun

Rain forests are hot because of the way the sun's rays strike the earth at the equator.

You can think about the sun and the earth by making a model like the one shown in the diagrams. The globe represents the earth and the flashlight represents a beam of sunlight.

A sunbeam carries heat as well as light so, away from the equator, the sunbeam's heat is spread over a large area. At the equator, the sunbeam heats a much smaller area. This means that the surface of the planet at the equator, where the rain forests are, gets much hotter than many places to the north and the south.

▲ When the beam strikes the earth where the surface curves sharply away from the sun, the beam covers a large area of the planet.

▲ At the equator, the surface of the earth curves only slightly away from the sun. The beam covers a much smaller area.

	Tropical rain forest	Temperate forest*
Day temperature	95° F (35°C)	86° F (30°C)
Night temperature	64° F (18°C)	50°F (-10°C)

*Forests in many parts of Europe, Northern Asia, and North America.

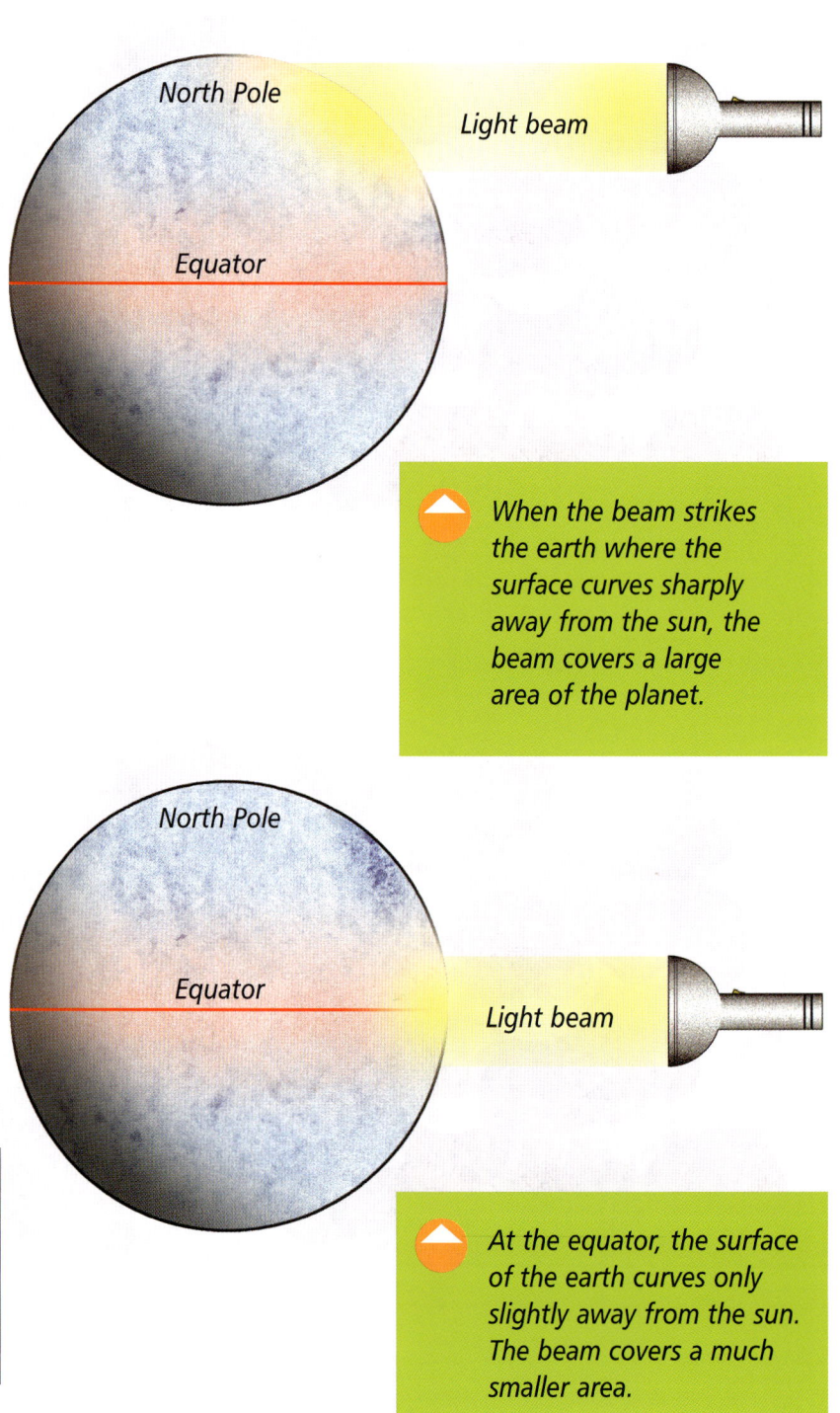

8

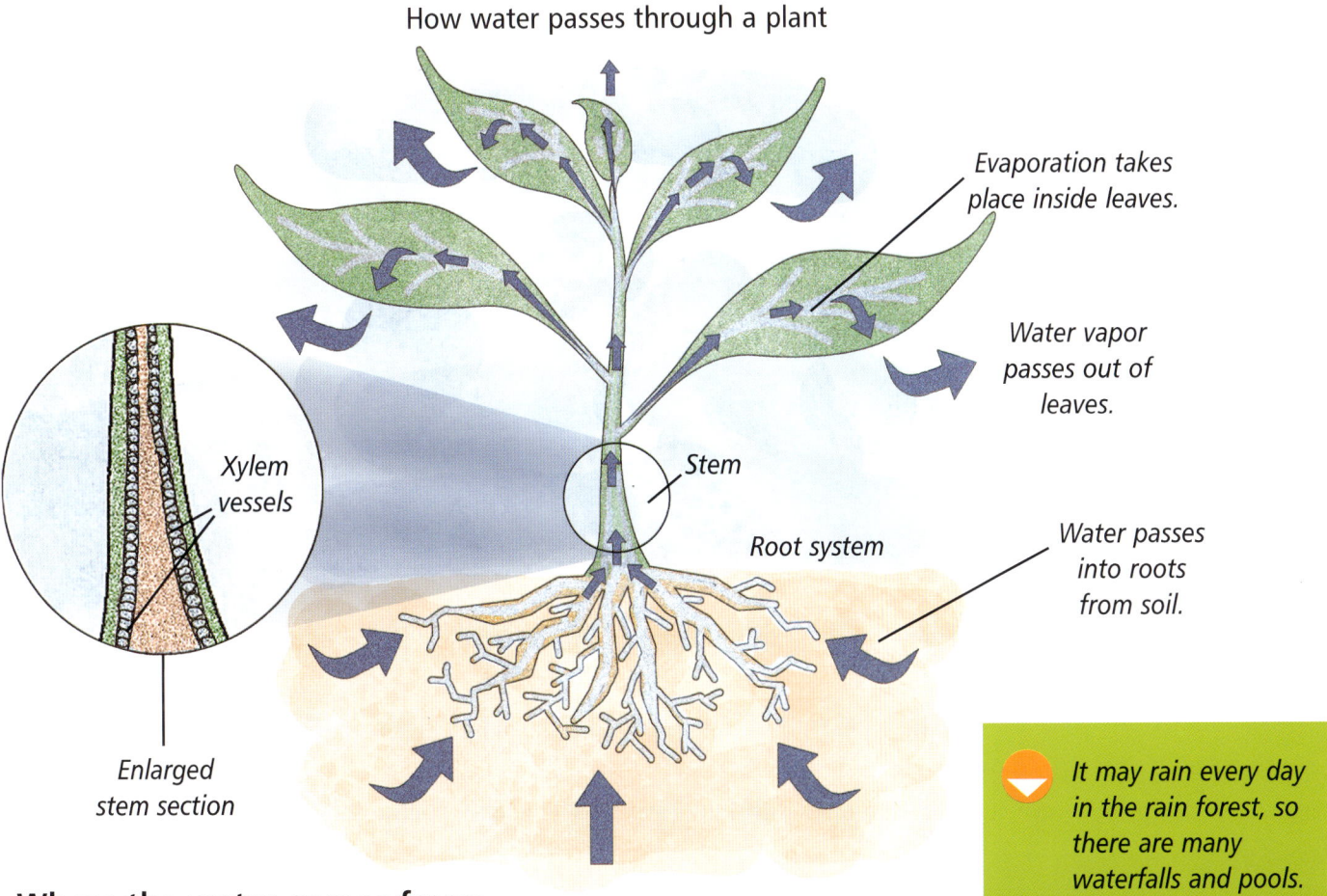

How water passes through a plant

- Evaporation takes place inside leaves.
- Water vapor passes out of leaves.
- Xylem vessels
- Stem
- Root system
- Water passes into roots from soil.
- Enlarged stem section

> It may rain every day in the rain forest, so there are many waterfalls and pools.

Where the water comes from: The rain forest water cycle

The large amounts of light and heat at the equator help many plants, especially trees, to grow to a huge size. But plants also need water to grow well. Rivers flowing through the rain forests provide some of the water that the plants need, but the plants also get another supply of water by making rain clouds.

Most plants take in water through their roots. The water passes through the plant along tiny tubes, called xylem vessels, that carry it into the leaves. Inside the leaves some of the water is used to make food, but much of the water changes to water vapor by a process called **evaporation**. The water vapor then passes out of the leaves through tiny holes in the leaf surface.

Around the outside of the leaves the warm air rises because of **convection currents.** These begin near a warm surface, such as a branch of leaves that the sun is shining on.

Air above the leaves warms up and becomes lighter in weight than the air around it. This makes the warm air rise through the cooler air and the cooler air moves in to take its place. As the warm air rises, it takes the water vapor with it.

The warm air continues to rise until it cools. This happens high above the rain forest. The water vapor in the cool air **condenses** on dust particles that are floating in the air and forms tiny water droplets. The droplets join together to form larger drops that form clouds and eventually fall as rain. Then the water is taken in by plant roots, starting its journey through the water cycle again.

Convection currents

Cool air falling

Warm air rising

Water vapor from leaves rises with warm current

> *Mist forms between the trees as water vapor rises from the soil and plants, cools, and condenses.*

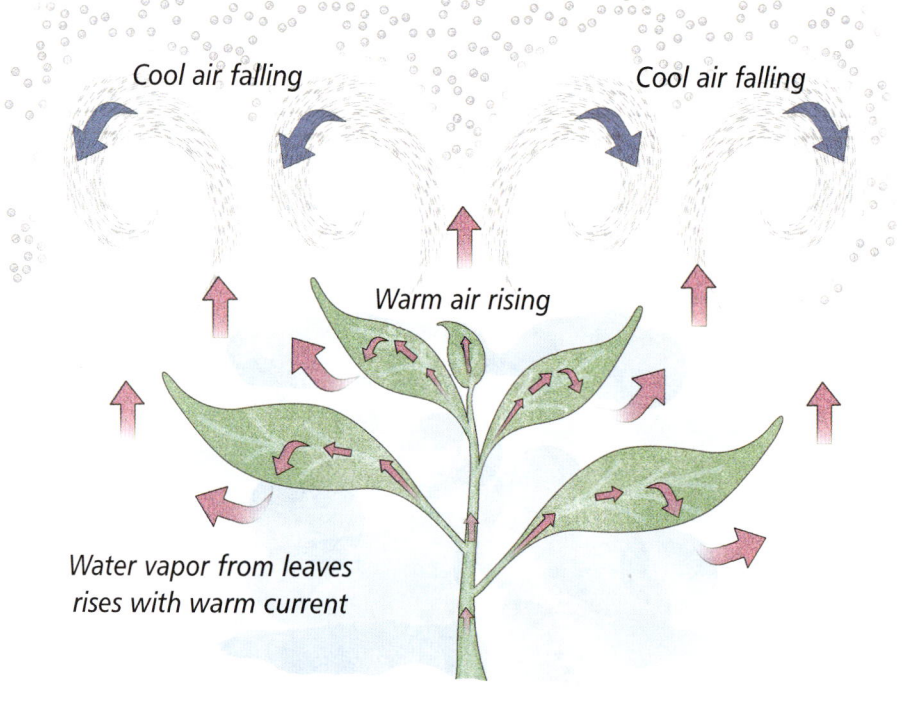

The rain forest water cycle

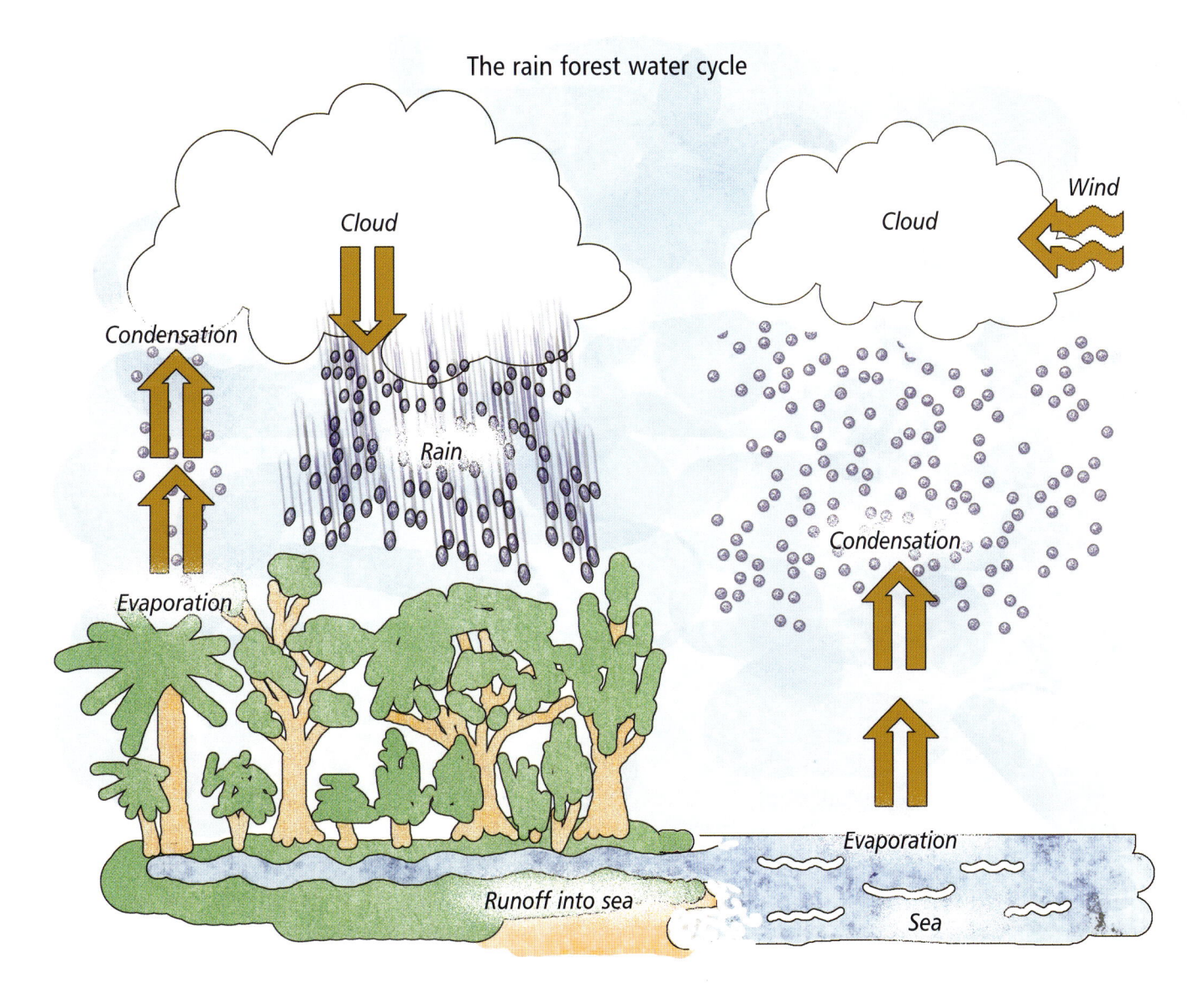

Diagram labels: Cloud, Condensation, Evaporation, Rain, Cloud, Wind, Condensation, Evaporation, Runoff into sea, Sea

At many times during the year it may rain every day in the rain forest. This is because there is no wind to blow the rain clouds away.

Not all of the rain that reaches the forest floor enters the plants. Some of it soaks through the soil and runs into streams and rivers. This water is carried away to the ocean, but it may return to the rain forest in the following way.

The heat from the sun's rays warms the surface of the ocean and makes some of the water change to water vapor. This rises on convection currents and condenses to form clouds, which may be blown back over the rain forest to fall as rain once more.

Humid air

As water evaporates from the plants and also from the surface of the soil, water vapor gathers in the air. This makes the air feel damp and clammy, like the air in a bathroom after someone has had a hot bath or shower. Air that feels like this is called **humid** air. Some of the water vapor may condense near the rain forest floor and form a mist between the trees.

Clothes for the Rain Forest

Rain forest people usually wear very few clothes. This helps them to keep cool. They do not have to worry about protecting their skin from the sun, because they live in the shade of the forest and the weather is often cloudy. They travel on well-worn paths, clear of thorny branches that would cut their skin.

Explorers, on the other hand, often need to make their own paths through the vegetation, cutting down branches with a sharp knife called a **machete,** or parang. They could easily be scratched or cut themselves, so they need clothing to protect their skin.

The clothing must have other **properties,** too. It must be light in weight, because heavy clothing would be exhausting to wear in the heat. The clothing must be capable of drying quickly, as very wet clothing becomes uncomfortable and can rot in rain forest conditions.

The Mbuti people live in the African rain forest. They build small, round huts from flexible tree shoots, covered with leaves. They set up camp near the edge of the forest, so that they can trade with other people who live outside the rain forest. Like many rain forest people, they decorate their skin with paint.

Day and night wear

Explorers have two sets of clothes. They wear one set as they travel through the forest during the day and carry the other set in a waterproof bag. At night they change out of the wet day clothes and put on the dry clothes for staying around the camp and for sleeping. They hang the wet clothes up overnight, so that they can dry as much as possible before morning.

Which cloth dries fastest?

This activity shows how the scientific method helps in choosing materials for making rain forest clothes. In Step 1 you are making an observation and having an idea. In Steps 2–5 you are doing an experiment to test the idea. Step 6 is your result and Step 7 is the conclusion.

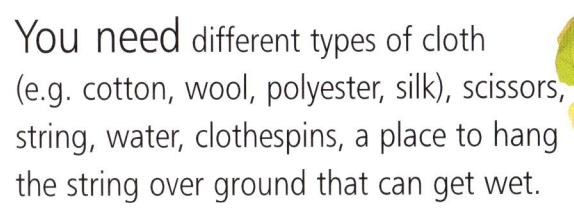

You need different types of cloth (e.g. cotton, wool, polyester, silk), scissors, string, water, clothespins, a place to hang the string over ground that can get wet.

1. Look at the types of cloth and decide which you think will dry fastest.

2. Cut a square from each material. All the squares should be the same size.

3. Set up some string like a clothesline.

4. Soak each square of material in water and pin it to the string.

5. Feel each material every hour to find out how fast they are drying.

6. Put the pieces of cloth in order, starting with the one that dries most quickly.

7. Which cloth would be best for making clothes for the rain forest?

Clothes for a rain forest explorer

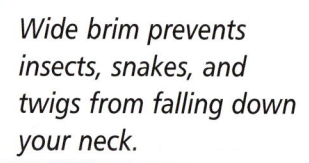

Wide brim prevents insects, snakes, and twigs from falling down your neck.

Loose-fitting shirt and pants allow air to move away from your body, carrying away the heat when you are hot and sweating.

Buttoned collar and cuffs keep out insects.

What to wear in the rain forest

Rain and thorns are not the only hazards you will meet if you travel through the rain forest. You may also be attacked by animals such as mosquitoes and leeches (see pages 38–39). Here is a set of clothes that will help you cope with rain forest conditions. They will also keep you cool by allowing your skin to sweat.

Tough material for shirt and pants resists tearing. Material is also quick drying, so it does not rot.

Plastic bag contains dry clothing.

Pants tucked into socks and boots prevent insects and rain from reaching the feet.

Boots cover the lower leg to prevent insect and snake bites.

Holes above the sole allow sweat from the feet to **evaporate.**

How sweat cools you down

The human body circulates its heat in the blood and is healthiest at a temperature of about 98.6° F (37° C). The body has some clever ways of keeping its temperature at this level. If its temperature rises, more blood is sent to blood vessels near the skin surface. Heat from the blood passes through the skin by **conduction** and then leaves the skin surface by **convection** and **radiation**.

If the body temperature still does not fall, the skin produces sweat. Heat from the skin then passes into the sweat by conduction. The sweat absorbs large amounts of heat, but the heat does not stay there. It makes the water in the sweat evaporate and change to water vapor. The water vapor carries the heat away and the skin cools down.

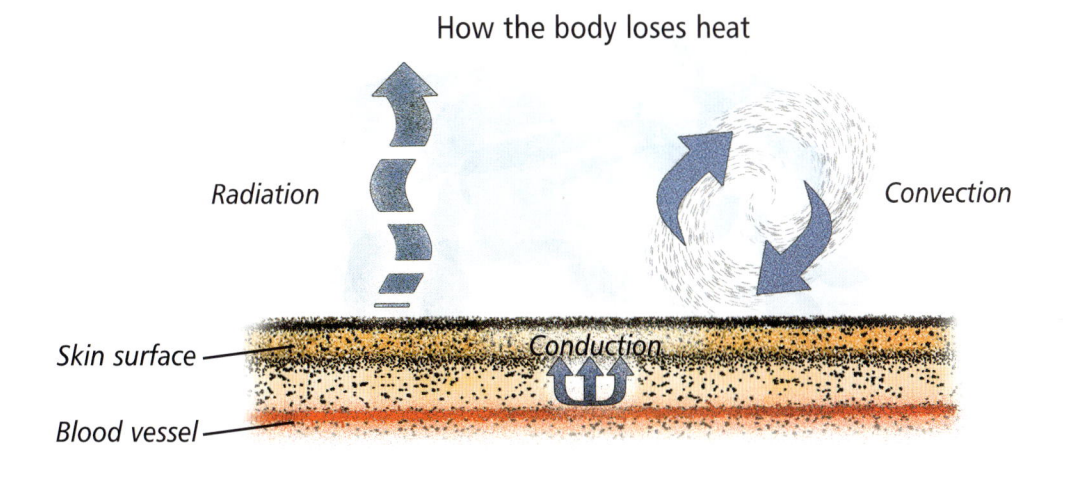

How the body loses heat

Radiation   Convection

Skin surface — Conduction
Blood vessel —

Why you need to wash your clothes

You will sweat a lot as you travel through the rain forest. Sweat is not just water. It also contains oils and these can sink into the fibers in your clothes. The oils are food for tiny living things called **bacteria.** In the heat and damp of the rain forest, bacteria breed quickly and soon an enormous number of them will be feeding on the clothes. If clothes are not washed every two days or so, they begin to rot as the bacteria eat them away.

These butterflies are drinking the sweat that has been absorbed by an explorer's sock. Sweat contains salt, too, and this is also useful for the butterflies.

Rain Forest Plants

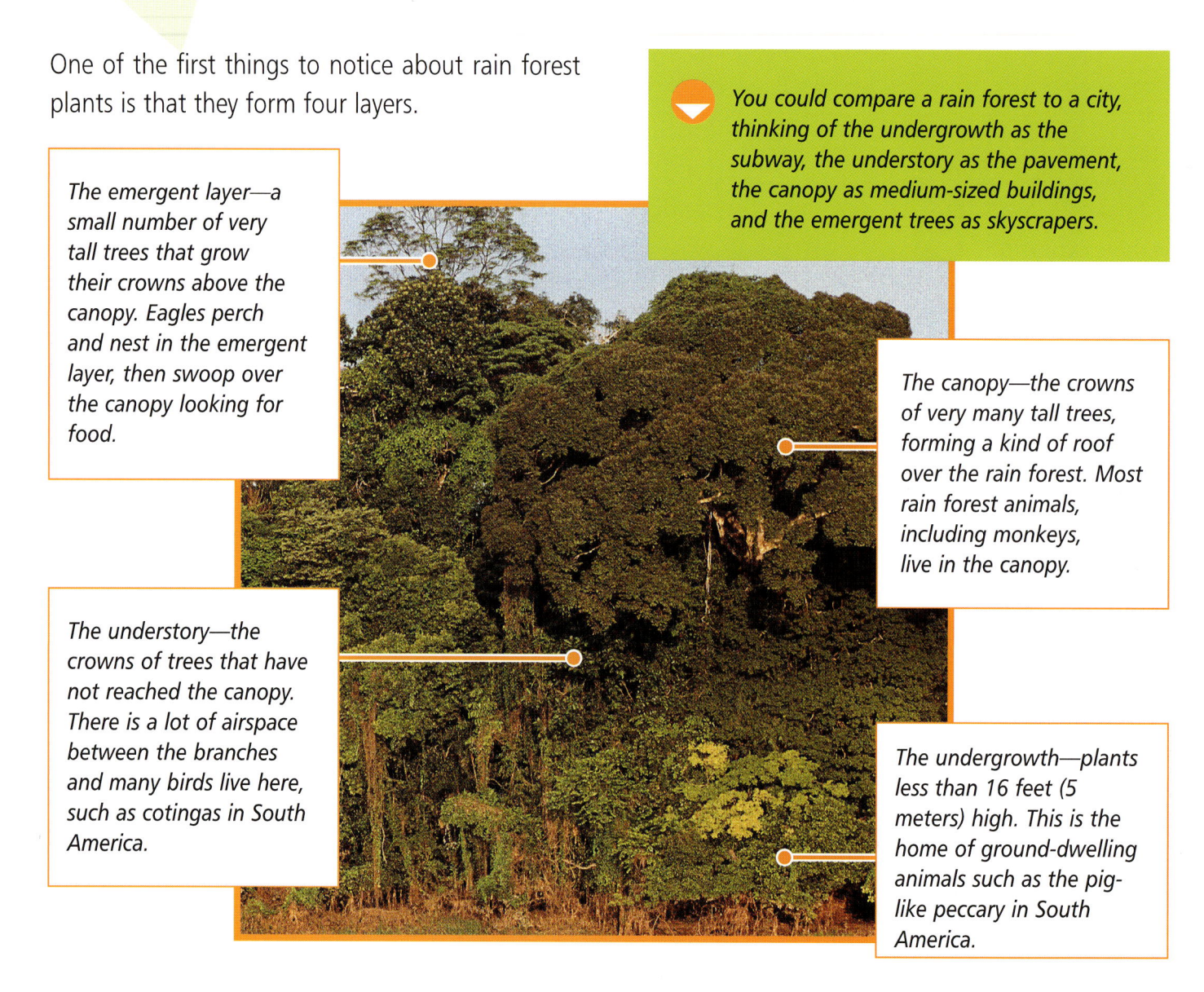

One of the first things to notice about rain forest plants is that they form four layers.

> You could compare a rain forest to a city, thinking of the undergrowth as the subway, the understory as the pavement, the canopy as medium-sized buildings, and the emergent trees as skyscrapers.

The emergent layer—a small number of very tall trees that grow their crowns above the canopy. Eagles perch and nest in the emergent layer, then swoop over the canopy looking for food.

The canopy—the crowns of very many tall trees, forming a kind of roof over the rain forest. Most rain forest animals, including monkeys, live in the canopy.

The understory—the crowns of trees that have not reached the canopy. There is a lot of airspace between the branches and many birds live here, such as cotingas in South America.

The undergrowth—plants less than 16 feet (5 meters) high. This is the home of ground-dwelling animals such as the pig-like peccary in South America.

Rain forest plants have all the water and warmth they need, but another key to their survival is light.

The canopy stops a large amount of light from reaching the plants beneath it and so these plants must either be adapted to living in dim light or they must grow fast. Plants like vines and **lianas** have thin woody stems that can grip a tree trunk. They grow quickly up the bark until they reach the canopy where there is more light.

Generally, there is too little light for much undergrowth to thrive, so the rain forest floor is somewhat bare of plants. A thick layer of undergrowth only forms where there is a gap in the canopy, such as by a river or in a clearing.

Do seeds grow better in rain forest conditions?

You can make rain forest conditions by putting a plastic jar over some soil. When sunlight shines on the jar, some heat is trapped inside. This makes the water in the top of the soil evaporate and form water vapor, so the air becomes humid. At night, when the air cools, the water vapor condenses on the sides of the jar and flows back into the soil. Compare how seeds grow in ordinary conditions and in rain forest conditions.

You need a packet of mustard or cress seeds, two flowerpots, soil, water, a plastic jar, a sunny windowsill.

1 Put the same amount of soil in each pot.

2 Plant the same number of seeds in each one.

3 Pour the same amount of water onto the soil in each pot to make it damp but not wet or waterlogged.

4 Put a plastic jar over the soil in one pot.

5 Put the two flowerpots in a sunny window and look at them every day.

6 Write down your observations every day.

7 What is your conclusion?

Tree trunk supports

Many rain forest trees grow so tall that they need extra support to stop them from falling over in a storm. In shallow soil some trees gain support by growing **buttress roots.** In wet soil some trees grow **prop roots.** The roots produce upward forces when the trees sway. These forces push against the weight of the trunk so the tree does not fall.

Plants without roots

Algae and moss are plants that don't have roots. They grow on tree bark, **absorbing** the water they need from the moist air.

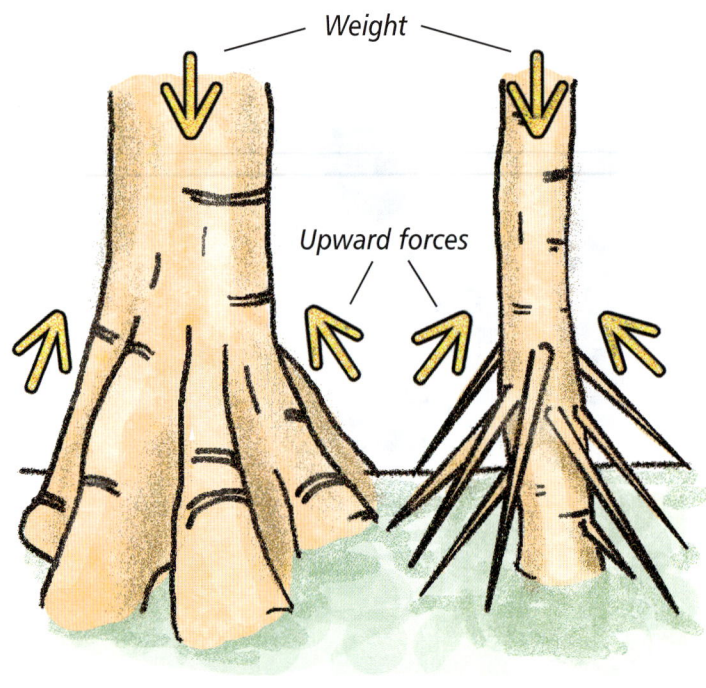

Buttress roots   Prop roots

Keeping leaves clean and dry

A plant must keep its leaves clean so that they can collect all the light they need. If the leaf surface is rough and pitted, water can collect on it and algae and moss can begin to grow. In time these plants would cover the leaf and no light would reach it.

The leaves of most rain forest plants have a smooth, thick covering of wax, which makes water run off them and stops other plants from growing on them. The leaves also come to a point called a drip tip, which helps the water flow away quickly. A leaf without a drip tip takes about five times as long to drain the water from its surface.

*The shape of the leaf funnels the water to the tip. The water forms a drop, which becomes so heavy that it is pulled by **gravity** to the ground.*

Traveling Through the Rain Forest

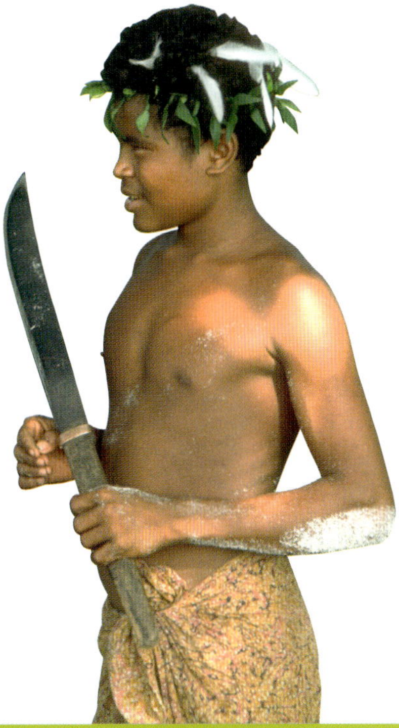

Cutting a way through

At the edge of a rain forest light can reach the ground, so shrubby plants grow close together and vines may grow between them. All these plants block your way into the forest unless you have a **machete** to cut your way through.

It's easier once you enter the dark space beneath the canopy, where few plants can grow on the forest floor. But occasionally, where there is a hole in the canopy, you will come across more plants that block your way and you will have to use the machete again.

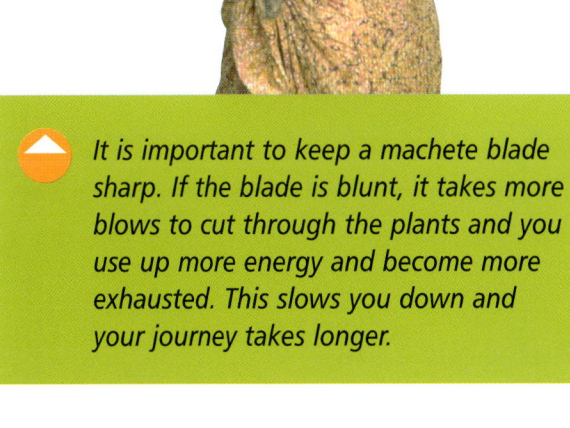

It is important to keep a machete blade sharp. If the blade is blunt, it takes more blows to cut through the plants and you use up more energy and become more exhausted. This slows you down and your journey takes longer.

Forces in lianas

Weight

When people walk across a rope bridge, their weight pushes down on the pieces of wood that make the walkway. This is balanced by the pulling forces that develop in the lianas. They stop the bridge from breaking and allow people to cross safely.

Crossing a ravine

Rain forest people make paths that become well worn so they can move around without always cutting their way through. To cross a **ravine** with steep sides, they use plants like **lianas** to make a bridge.

Lianas have very long stems, which grow up the sides of trees and across the canopy. Some stems hang down, and monkeys use them to swing from tree to tree. Lianas have very good tensile strength, which means that they can be pulled hard without snapping. This makes lianas useful for making a rope bridge.

Travel by boat

You do not go very far in a rain forest before you come to a river. Some of the many rivers are small, formed from streams. They join to make larger, wider rivers, which take the rainwater to the ocean.

On an **expedition** there may be many people and a large amount of equipment, such as tents, cameras and food containers. It is easier to carry all this equipment by boat on a river than it is to carry it on your back as you walk. So many expeditions use boats as a means of transport wherever possible.

A dugout canoe

Rain forest people also use boats on the river to carry goods between villages and to fish. They use a boat called a dugout canoe, which is made by cutting down a large tree and hollowing out the trunk using axes. When most of the wood has been removed from the trunk, fires are lit inside it. The heat does not set the canoe on fire but causes carbon, the black substance that forms when wood is burned, to harden the wood. It also seals any gaps in the wood, making the canoe watertight.

These people in Papua New Guinea are hollowing out a tree trunk to make a canoe.

Make a model dugout canoe

Wood is difficult to carve, but modeling clay is much easier. A lump of modeling clay sinks in water because its weight is greater than the upward pushing force of the water, called the **buoyancy.** Making a hollow in the modeling clay reduces its weight. The air that fills the hollow weighs very little. You should be able to make a model dugout canoe that pushes down on the water with a smaller force than the buoyancy and therefore stays afloat.

You need a large piece of modeling clay, a plastic knife, a bowl of water.

1. Roll some modeling clay into a thick cylinder shape.

2. Put the cylinder in water and watch it sink.

3. Take it out of the water and use a plastic knife to hollow out some of the modeling clay.

4. Put the hollow cylinder back in the water. If it sinks, take it out of the water and cut away some more modeling clay.

5. Keep testing and cutting until you make a canoe that floats.

Finding a Way

When you see tree trunks all around you, and every direction looks the same, it can be very hard to find your way.

When you look around in the rain forest, each direction looks the same. This is because of the huge number of trees. All their trunks grow up into the canopy and all look similar. This can have a frightening effect and, after a short time, it can make you lose your sense of direction so that you are in danger of going around in circles, even though you think you are moving in a straight line. Many rain forest explorers have died because they lost their sense of direction.

In some environments where the sky is clear, the position of the sun can help you find directions. This is because the sun always rises in the east and sets in the west. However, you cannot find directions this way in the rain forest, because the sun is hidden by the canopy. Instead, you could use a compass to tell you the position of North. A compass pointer is a magnet that is allowed to move freely. It always comes to rest pointing North. To understand why, you need to know about Earth's **magnetism**.

Warning!
Never look directly at the Sun.
It can damage your eyes.

The "magnet" inside Earth

At the center of the earth is a huge ball of iron and nickel, surrounded by a liquid layer of the same metals. As the earth spins, the ball moves in the liquid and this movement is believed to make the inside of the earth behave as if it contained a huge bar magnet. One end of this "magnet" in the earth points close to the North Pole and the other points to the South Pole.

A cross section of the earth

Crust (solid rock)

Liquid layer

Solid metal core

Mantle (partly molten rock)

Imaginary magnet

▲ The same poles of magnets repel each other.

▲ Opposite poles of magnets attract each other.

The poles of a magnet

All magnets have a north pole and a south pole and, if you bring two magnets close together, you can see how the poles make the magnets behave. If you bring the two north poles or the two south poles of the magnets together, they push each other apart, because similar poles repel each other. If you bring the north pole of one magnet close to the south pole of the other magnet, they pull together or attract each other.

The magnet inside the earth is so powerful that its poles repel and attract the poles of other magnets without having to get close to them. If any magnet is allowed to move freely (for example, by hanging it from a thread), its poles are attracted and repelled by the poles of the earth's "magnet" so that it comes to rest pointing in a north-south direction.

Magnetic materials

Some materials are magnetic and others are nonmagnetic. A nonmagnetic material is not attracted to a magnet. A magnetic material is one that is attracted to a magnet. Also, a magnetic material can be made into a magnet. The most common magnetic materials are iron and steel. Almost all other materials are nonmagnetic.

> Test some everyday objects to find out which materials are magnetic.

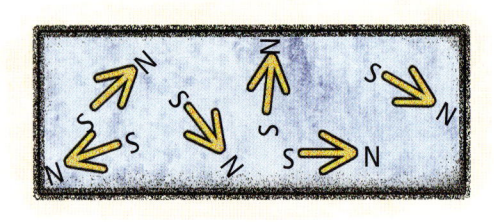

Domains before nail is magnetized.

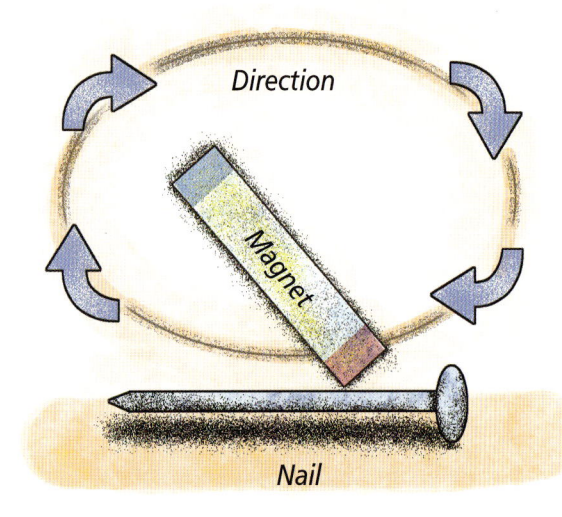

Stroking nail with magnet.

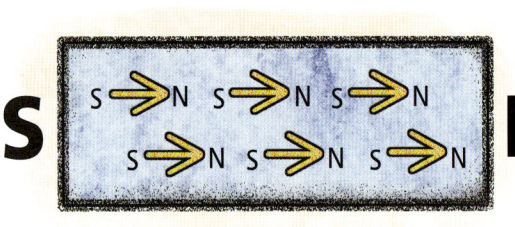

Direction

Magnet

Nail

Domains in magnetized nail.

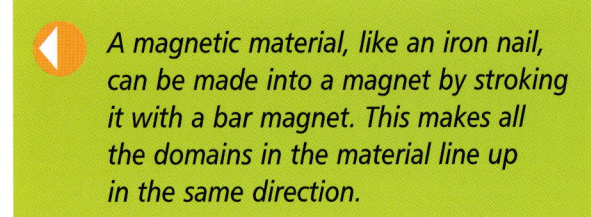

How a magnet is made

All materials are made up from tiny particles that can only be seen with very powerful electron microscopes. In iron and steel, groups of these particles arrange themselves together into microscopic magnets called domains. Normally these microscopic magnets point in all directions. However, they can all be made to point in the same direction if the iron or steel is stroked with one end of a magnet.

When the domains are all lined up in the same direction, one end of the piece of metal has a north pole and the other end has a south pole. It has become a magnet.

> A magnetic material, like an iron nail, can be made into a magnet by stroking it with a bar magnet. This makes all the domains in the material line up in the same direction.

Make a compass

In the gloom of a rain forest it is very difficult to figure out directions. A compass shows North and South, and you can use these two directions to work out others.

You need a needle, a magnet, a small piece of wood, a bowl of water.

1. Hold down the needle and stroke it with a magnet about 40 times. Always move the magnet in the same direction, starting at the same end of the needle. Stroking the needle in this way makes it become a magnet.

2. Place the needle magnet on the wood and float the wood in the center of the water. Do not let the wood touch the side of the bowl.

3. To find out which end of your needle magnet is its north pole, bring the north pole of a magnet close to one end of the needle. If the end of the needle turns away, that end is its north pole. If the end of the needle moves towards the magnet, then the opposite end of the needle is its North pole (see poles, page 23).

4. Float the needle on the water again without the magnet. It may move around a little and then settle. Note the direction of the pole end of the needle. It is pointing north.

Making a Shelter

In the rain forest there are materials all around that can be used to make a shelter.

Building the framework

The framework is made from wooden poles, which are obtained by cutting down small trees called saplings with trunks about 3.5 inches (9 centimeters) thick. Four poles are driven into the ground to make the corners of the shelter and then the other poles are attached to them.

Nails for fixing poles together would be too heavy to carry on an **expedition,** and there is not time to make joints in the wood. Instead, the poles are lashed (tied) together with lengths of vine. Vines grow up the sides of trees. They must be pulled down carefully, to avoid bringing dead branches, snakes, or spiders crashing to the forest floor. The vine is tied to one pole with a hitch knot and then wrapped or "lashed" around this pole and another to join them together securely. The lashing is completed by making two more hitch knots around the poles.

This typical rain forest shelter is raised off the ground and the walls are made of leaves.

The framework of a shelter

Sloping roof will be covered with leaves.

Area for bed.
Bed is off the wet ground.

In a lashing the vine is wrapped around an upright pole and a horizontal pole to join them together.

Tie a timber hitch knot

A knot is made by pushing and pulling a rope or string in a certain way. The timber hitch knot is used to connect one end of a rope to a piece of wood to start a lashing.

You need an 8-inch (20-centimeter) piece of string and a pencil.

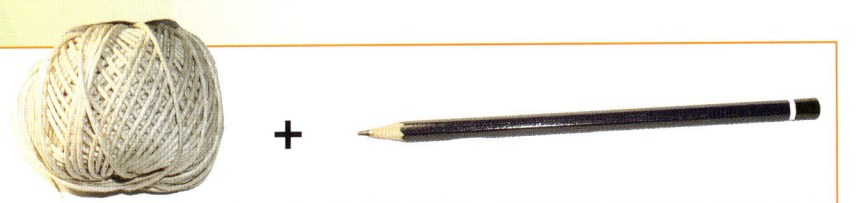

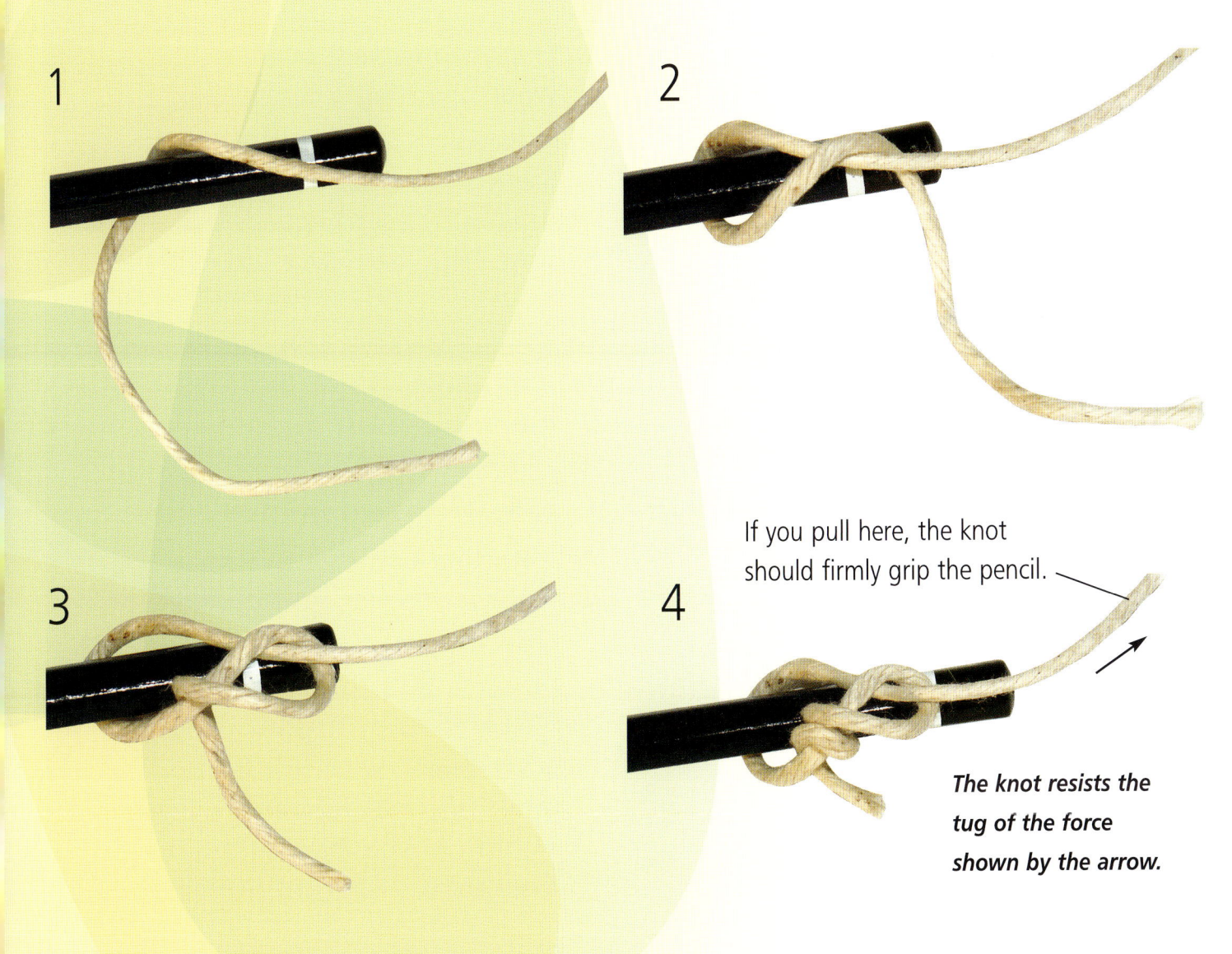

1

2

3

4

If you pull here, the knot should firmly grip the pencil.

The knot resists the tug of the force shown by the arrow.

27

A large waxy leaf can give some protection from the rain.

Making the walls and roof

The framework of poles needs to be covered with a material that will make the shelter waterproof. Explorers may use cloth, but rain forest people use leaves. Many plants that grow near the ground have huge leaves that can easily be cut down. These leaves are often made up from smaller parts called leaflets, which grow out from the central stalk, so the whole huge leaf looks like a giant feather. When leaves like these are put together, the leaflets can be arranged to overlap each other, like tiles on a roof.

The leaves need to be held in place with thinner cord than vines. This is made from fibers taken from the stems or leaves of plants. Rain forest people make holes in the leaves and push the cord through to hold them together.

A place to sleep

Many kinds of animals, such as scorpions and snakes, are active on the forest floor at night. It's important that shelters where people rest and sleep are raised off the ground. Poles, vines, thin branches, and leaves are used to make beds.

Things to beware of when choosing a campsite

- **Paths**

Some large animals that live on the forest floor, such as jaguars, leopards, and wild pigs, make paths through the undergrowth and use them regularly. To avoid these nighttime visitors, set up camp away from such paths.

- **Rotting logs**

Clear the ground of small rotting logs and other dead vegetation. These are sometimes home to snakes and large insects.

- **Rotting branches overhead**

If a storm develops, rotting branches could break off from the trees above and fall onto the shelter. They could possibly injure people resting inside.

- **Streams**

Set up camp away from streams. If the rains are heavy enough, the stream may turn into a river and wash the camp away.

Extracting plant fibers

You can see how fibers can be taken out of plants by crushing a piece of celery.

You need a long celery stalk, a piece of wood or a breadboard, a rolling pin.

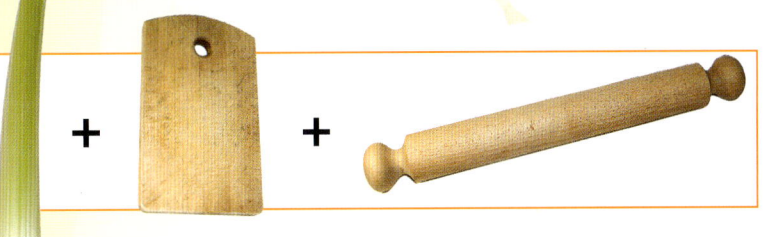

1 Lay the celery stalk on the piece of wood.

2 Squash the celery by pressing down on it as you roll it with the rolling pin. Part of the stalk will become white and mushy, while in the other part you should see fibers running along the length of the stalk.

3 Pull out the soft white part of the celery to see the long fibers better.

4 Pull out the fibers and try to get some that are the full length of the stalk.

5 Pull the fibers to test their **elasticity** and their strength.

6 Try to tie a timber hitch knot in one fiber (see top picture and page 27 for how to do it).

Building a Campfire

When people make a camp in a rain forest, it is not long before they get a fire going to boil water, cook their food, and dry their clothes. A campfire helps people feel more comfortable and at ease. It also keeps large, dangerous animals such as jaguars away.

Making a fire

In the rain forest, it is sometimes difficult to find dry material to make a fire. People often carry some small pieces of dry material, called **tinder,** that are used to get a fire going. Tinder can be made from dry stalks of grass, dry moss, or even dry hair.

One way to make a fire in damp conditions is to use a fire drill. This consists of a thin wooden pole (called the "drill"), which is made to spin back and forth very quickly inside a hole in a flat piece of wood (called the "hearth").

Friction between the drill and the hearth causes so much heat that any tinder close by bursts into flame. Small sticks can then be placed over the tinder to build up the fire.

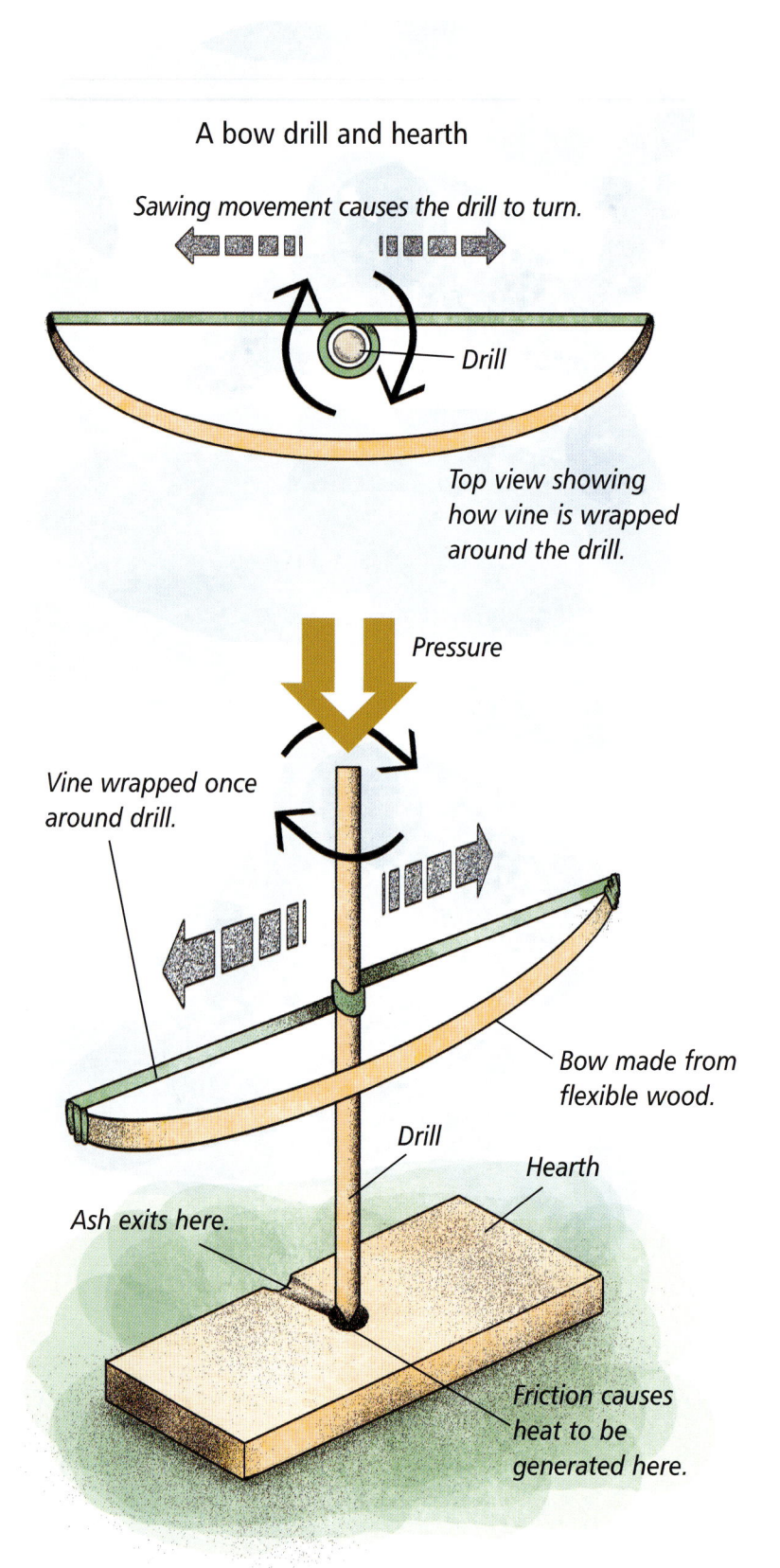

A bow drill and hearth

Sawing movement causes the drill to turn.

Drill

Top view showing how vine is wrapped around the drill.

Pressure

Vine wrapped once around drill.

Bow made from flexible wood.

Drill

Hearth

Ash exits here.

Friction causes heat to be generated here.

30

> *Sitting by a campfire at night helps explorers relax. The fire also gives them protection from large and small animals.*

Smoke and ash

When something burns, carbon joins with oxygen in the air and heat is given out. Invisible gases, including carbon dioxide, are produced. Some of the carbon forms tiny black particles that float in the air and form smoke. The solid remains left after the fire are called ash.

Both smoke and ash help to keep insects away. Flying insects are driven away by smoke. One such insect is the mosquito, which can carry the disease **malaria**. This means that a fire can help prevent campers from being bitten by malaria-carrying mosquitoes.

Substances in ash drive away insects that crawl on the ground. Spreading cold ash under beds keeps these insects away from campers at night.

Carrying fire

Some rain forest people, like the Mbayaka Pygmies of Africa, take some of the embers from their campfire, wrap them in heat-resistant leaves, and carry them to their next camp. Here they use the embers to make a new fire. Carrying fire in this way makes it easier to light a new fire when a camp is made.

Water to Drink

At 212° F (100° C) water boils and changes into a hot transparent gas called steam.

Water vapor condenses on dust in air to make steam.

Long branch

Suspended can of water

Vine "rope"

Rocks holding down branch

Forked branch

Heat passes through base of can by **conduction**.

Convection currents *carry heat through water.*

▲ *Rain forest materials can be used to hang a can of water over a fire. All water must be boiled for ten minutes to make it safe.*

Water splashes through the canopy almost every day. You cannot travel far without coming to a stream or pond. Yet, although there is plenty of water, most of it is not safe to drink. As water flows over leaves or through the soil into the streams, it picks up tiny living things called **bacteria**. Some of these bacteria cause deadly diseases. However, if the water is boiled for at least ten minutes, the bacteria in it are killed and it is safer to drink.

Some people on **expeditions** carry tablets for purifying water. When the tablets are added to water, they kill bacteria.

There are many ways to collect water in the rain forest, but the water should always be treated to kill bacteria before it is ready to drink.

Collecting rainwater

A waterproof sheet can be tied between tree trunks so that, when it rains, water collects in the center. The water can then be poured into a pan for boiling. One shower may provide a few gallons of water.

Stream or river water

Water from a stream or river may have mud, leaves, and even small animals in it. These can be removed by filtering. A filter is a material with very small holes in it. Water can pass through the holes, but larger objects cannot. Filtered water must be boiled, because bacteria are small enough to pass through the holes with the water.

Can you make a filter from a sock?

In the rain forest, explorers sometimes have to **improvise,** or use something in an unusual way, to survive. Sometimes they must use articles of clothing, such as a shirt, tights, or a sock as a filter.

You need
an old sock, grass, sand, jar of water, leaves, soil, empty jar.

1. Line the sock with grass and put some sand on top of the grass.

2. Put some leaves and soil in the jar of water and stir it up.

3. Hold the sock over the empty jar and carefully pour some of the dirty water into the sock. Save some of the dirty water to make a comparison.

4. Let the water drain through the sock into the jar.

5. Compare the filtered water with the unfiltered water. How much of the solid material did your filter remove?

Pour mix in here.

Filtered water collects here.

Warning!
Do not drink the water. It could contain bacteria, which could harm you if you drank it.

Taking water from plants

On page 9 you saw how plants take in water through their roots and how water travels through a plant. Rain forest people and people on **expeditions** sometimes take water from plants, including some types of vine.

Two cuts are made in the vine stem. One is high on the stem, to cut the leaves from the stem and stop them from sucking up the water. The other cut is low, to let the water pour out of the stem. If a person is sure that the vine will give healthy water, he or she can hold it and allow the water to drip into their mouth.

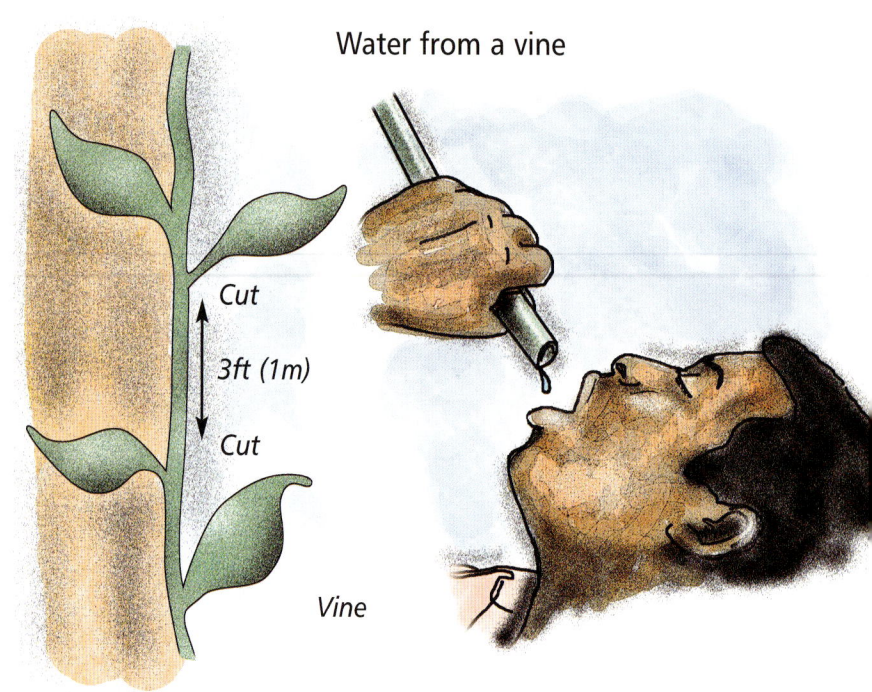

Water from a vine

Cut
3ft (1m)
Cut

Tree trunk
Vine

Warning!
Do not drink or eat from any plant unless you are sure it is safe.

In the Amazon rain forest, plants called **bromeliads** live on the branches of trees. Bromeliads do not have roots. Instead, their leaves grow so that there is a space in the middle where rainwater collects, and the plant takes the water it needs from this. If you tip the water out of a bromeliad, you may find **tadpoles** swimming in it—so filtering and boiling are necessary to make the water safe to use.

Rainwater has collected in the "bowl" made by the leaves of this bromeliad. Bromeliads grow on tree branches, and some are so big that small animals like frogs live in the water that collects in them.

In Southeast Asia there are pitcher plants. The ends of their leaves form containers, called "pitchers," which fill with rainwater. If this water is emptied out, a lot of dead insects may be found in it. The water would have to be filtered and boiled to make it safe for drinking.

The importance of salt

Your body needs salt to make the nerves and muscles work properly. When you sweat, salt leaves the body—perhaps you have tasted the salt in sweat that has run down your face. If your body loses large amounts of salt and it is not replaced, you become tired and dizzy, feel sick, and develop muscle cramps.

At home, a normal diet gives you enough salt to meet your needs. In the rain forest you may be eating a diet with less salt in it, and you may be sweating much more than usual, and so you may lose more salt than you take in. People on expeditions carry salt tablets. A tablet is broken up and one piece is dissolved in plenty of water. Drinking this replaces the salt their body has lost.

On expeditions in Southeast Asia, people who run out of salt tablets can dig up the roots of the Nipa palm and boil them. When all the water has boiled away, salt crystals are left behind.

A pitcher plant traps insects in the water that collects inside it. Chemicals from the plant break down the bodies of the dead insects, so that the plant can absorb them as food.

Food

A Penan chief hunts with a blowpipe in Sarawak, Southeast Asia.

Rain forest people hunt animals and gather plants for food. They are "hunter-gatherers." This is the way humans obtained food before civilization.

Catching a meal

Rivers running through the rain forest contain large numbers of fish. These may be caught with nets or spears or by poisoning the water with the juice from certain plants.

Animals that live on the forest floor or in the low branches of the undergrowth are hunted with bows and arrows. Animals that live high in the trees are hunted with a blowpipe and poison darts. Poison from plants or from the skin of certain types of frogs is rubbed onto the tip of the dart. The dart is placed in one end of the blowpipe.

When the hunter sees an animal, such as a monkey or a parrot, he aims the blowpipe at it and blows into the pipe. The force of the air pushes the dart out of the pipe at great speed. If the dart strikes the animal, the tip pierces its skin and the poison enters its blood. The poison kills the animal and it

The sago palm is the main source of food starch for the Penan people.

falls to the ground. The meat from the animal is then cooked to destroy the poison so that the food is safe to eat.

Food from plants

Although a rain forest is full of plants, many of them cannot be eaten by humans and some are deadly poisonous. Even if one part of a plant is edible, it does not mean that other parts can be eaten too. For example, the tomato is a rain forest plant from South America. It produces edible red fruits, but the rest of the plant is poisonous.

Many foods from plants can be eaten raw, but some must be cooked. The mango and the pawpaw are examples of fruits that can be eaten raw. They are also sold in shops throughout the world, so perhaps you have eaten them before.

Young shoots of the bamboo plant can be cooked to remove their bitter taste, and the seeds of the bamboo can be cooked like rice. Sago palm trees grow in Southeast Asia. When a sago palm is cut down, its pith can be removed and cooked to make a thick, pale liquid food called sago.

Some plants have their food hidden from view. For example, the wild yam is a creeping plant whose roots are swollen with food, forming tubers. The tubers can be dug up, peeled, and boiled before eating, to remove any poisons they may contain.

In the Amazon rain forest, fruits may be collected and carried home in a basket woven out of palm leaves.

Dangers to Health

There are many dangers to health in the rain forest. You could drink water contaminated with harmful **bacteria,** eat a poisonous plant, or not get enough salt in your food (see page 35). You could also be attacked by animals. You may be surprised that the animals that attack most frequently are small. Large animals like leopards and snakes tend to hide away, but some small animals are out for your blood.

Mosquitoes

There are many kinds of mosquitoes. Some feed on the juices inside plants, but some need blood. A mosquito has a long thin tube for a mouth, like a needle used for injections. When the mosquito lands on the skin, it plunges the end of the tube into it. The mosquito then sends **saliva** down the tube, to stop blood clotting. This makes the blood easier to drink.

Germs enter the body in the mosquito's saliva. They mix with the blood and move through the body to the liver. Here they set up home and breed. When a huge number of these germs build up in the body, they cause a disease called **malaria.** Some people die from this disease. If another mosquito feeds on a person who has malaria, it picks up more of the germs and these live in the mosquito for a while before they leave in its saliva to infect another person.

When a mosquito bites, it sends its saliva into your blood. The saliva may contain malaria germs.

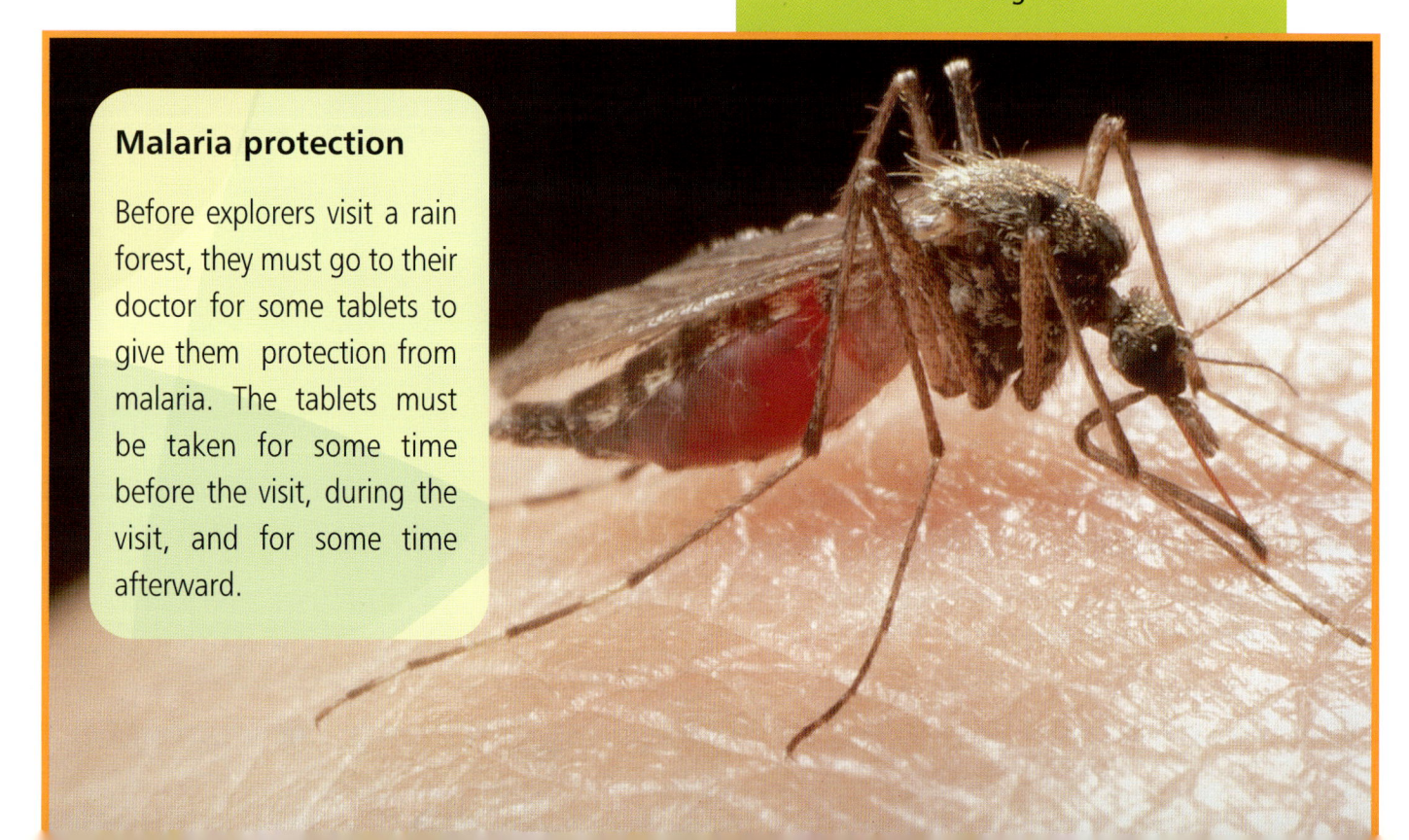

Malaria protection

Before explorers visit a rain forest, they must go to their doctor for some tablets to give them protection from malaria. The tablets must be taken for some time before the visit, during the visit, and for some time afterward.

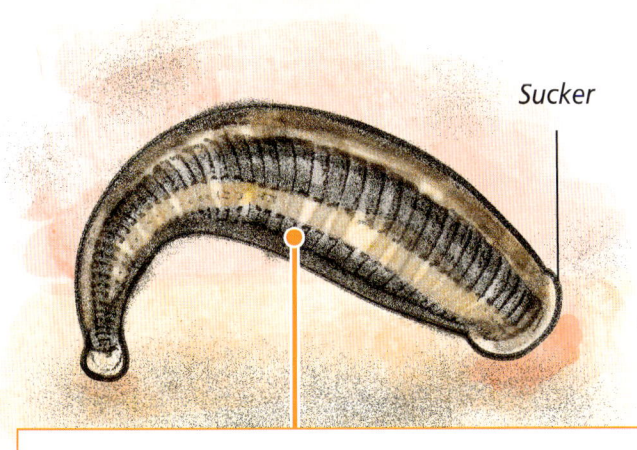

Sucker

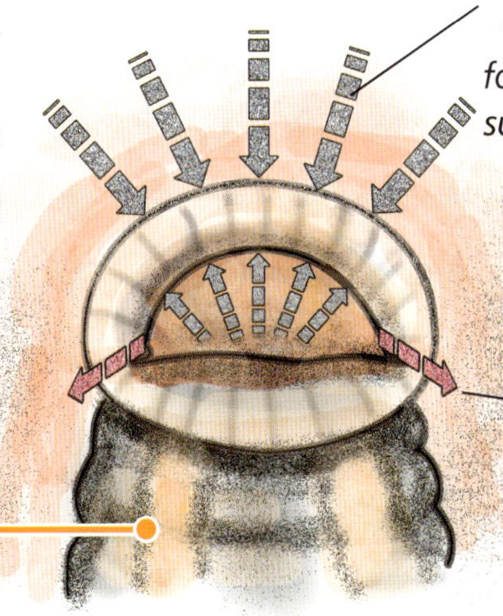

How a sucker works

Remaining air inside sucker pushes with a small force.

Air above sucker pushes with a larger force and holds sucker in place.

Some air is pushed out when sucker is attached. (pink arrows)

Leeches

A leech is a wormlike creature that uses its sucker to grip onto a leaf. When someone passes by, it drops off the leaf and uses its sucker to grip their skin. The leech bites to reach the person's blood, on which it feeds. It does not cause harm by sucking blood, but the wound left in the skin can become infected with germs and can make the person ill.

Legs have claws to hold on tight.

Ticks

A tick looks like a tiny spider. It lies in wait on a plant. When someone brushes past, the tick climbs onto his or her skin, holds on with its legs, and bites. The tick sucks out blood until it swells up and falls to the ground. Ticks cause harm in two ways: they put germs into the blood with their saliva, and the wound they leave behind can become infected with other germs.

Large compartment in abdomen to store blood.

Legs have claws to hold on tight.

Lice

Lice are insects without wings. They crawl into the clothing and hair of people in the rain forest and feed by biting the skin and sucking blood. They also leave their droppings on the skin. The droppings of rain forest lice are infected by a germ that causes a dangerous disease called typhus. People who are bitten by lice should not scratch their wounds as they may rub lice droppings into them and become infected by the typhus germs.

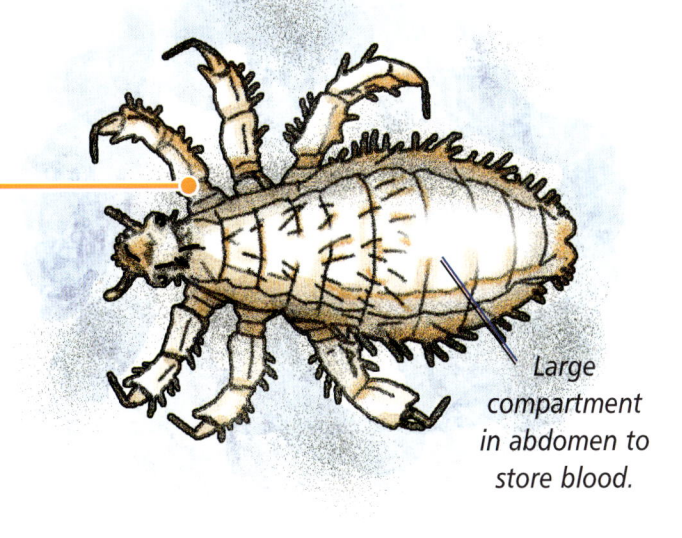

Large compartment in abdomen to store blood.

Rain Forest Animals

Some monkeys use their tails as a fifth limb to help them move through the trees.

When this frog calls, it blows up a fold of skin into a balloon shape. This forms a resonator, which makes the call louder.

The purpose of many rain forest **expeditions** is to study the animal life. More species of animals live in an acre of rain forest than in an acre of any other habitat, and many new species are discovered every year. Most animals live in the canopy and so, to study them closely, scientists climb into the tall rain forest trees using ropes like those used by mountain climbers.

Communicating

Most animals need to communicate with others of their own kind. They may do this by making themselves visible or by letting out sounds. In the rain forest canopy, the mass of leaves makes it difficult for animals to see each other, so many of them are brightly colored to increase the chance of being seen. The leaves also muffle sounds, so the animals have developed loud calls so that they can be heard over great distances.

Moving through the canopy

Many animals in the canopy are adapted to move through it with ease. Monkeys have long legs to help them jump from branch to branch. The colobus monkey also uses its long tail to keep it balanced as it leaps. Some monkeys, such as the spider monkey, have a **prehensile** tail. They coil the tail around a branch and hang from it to gather fruit. The tail acts as a fifth limb. Apes, such as the gibbon or orangutan, have very long arms that help them swing from one branch to the next.

Make a simple resonator

Sounds are made when an object vibrates. Frogs and mammals have a pair of flaps in their throat called vocal cords. They make these vibrate in order to make a call. Frogs and howler monkeys have resonating chambers that vibrate with the vocal cords to make the calls louder. You can demonstrate how a sound is increased in this way.

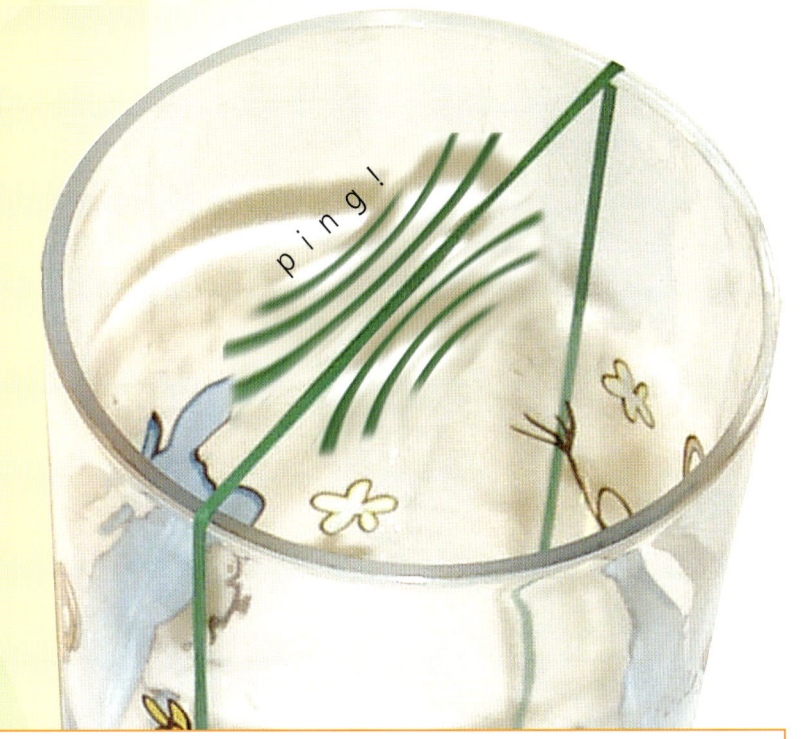

You need a rubber band, a small glass, eye protection. The rubber band should be large enough to fit over the glass without stretching too much.

1. Stretch the rubber band enough so that it will fit over the glass. Then pluck it and listen to the sound.

2. Put the rubber band around the glass and pluck the part of the band that is over the mouth of the glass. You should find that the glass acts as a **resonator** and makes the sound louder.

3. Ask a friend to help you check how the glass makes the sound louder. Make sounds as in step 1 and let your friend move away until he or she cannot hear them. Measure the distance between you and your friend. Then make sounds as in step 2 and use the same method to find out how far away they can be heard. You should find that the sounds made with the resonator can be heard at a greater distance.

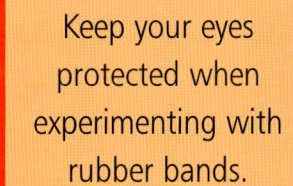

Pluck here.

Keep your eyes protected when experimenting with rubber bands.

Pluck here.

When giant flying squirrels leap through the air, they spread their skin folds like parachutes.

Gliding animals

Between the canopy and the understory is a large airspace where many birds fly and some animals travel by gliding. The gliding frog has large webs between its toes, so that each foot acts like a parachute as the frog makes a leap. The gliding lizard has extra bones in its ribs that stick out along the side of its body. There is skin between these bones that helps the lizard glide. When the flying snake throws itself into the air, its body becomes flat like a ribbon so that it can move through the air easily.

The most well-known gliding animals are flying squirrels. Folds of skin between their front legs and back legs act like parachutes as the squirrel moves through the air.

Rain forest snakes

A snake's long, limbless body can coil around branches and climb through trees with ease. Most snakes have sharp teeth to grip and kill their prey, but only a few kinds of rain forest snakes are poisonous. They use their poison when they bite, to kill their prey quickly and keep it from struggling. They also use it to defend themselves if attacked by another animal.

The largest rain forest snakes are the anaconda and the boa constrictor. These snakes are not poisonous. They kill by coiling their body around their prey to stop it from breathing. Snakes that kill in this way are known as constrictors and eat a wide range of animals. The largest snakes can kill wild pigs. A snake has extra bones in its jaws. They let the snake open its mouth very wide and swallow its prey whole.

How do skin folds affect a flying squirrel?

Scientists sometimes make models to test their ideas. In this activity you can make two model squirrels to test the question.

You need six pipe cleaners, scissors, tissue paper, sticky tape.

1. Make a model squirrel from three pipe cleaners.

2. Cut out two pieces of tissue paper and stick them between the front and back legs.

3. Make a second squirrel, but do not give it skin folds.

4. Hold up a squirrel in each hand, then let them go.

When an object falls, it is pulled down by the force of gravity. At the same time, the force of air resistance pushes upward on the object's surface and slows the speed at which it is falling. You should find that the squirrel with model skin folds has more air resistance.

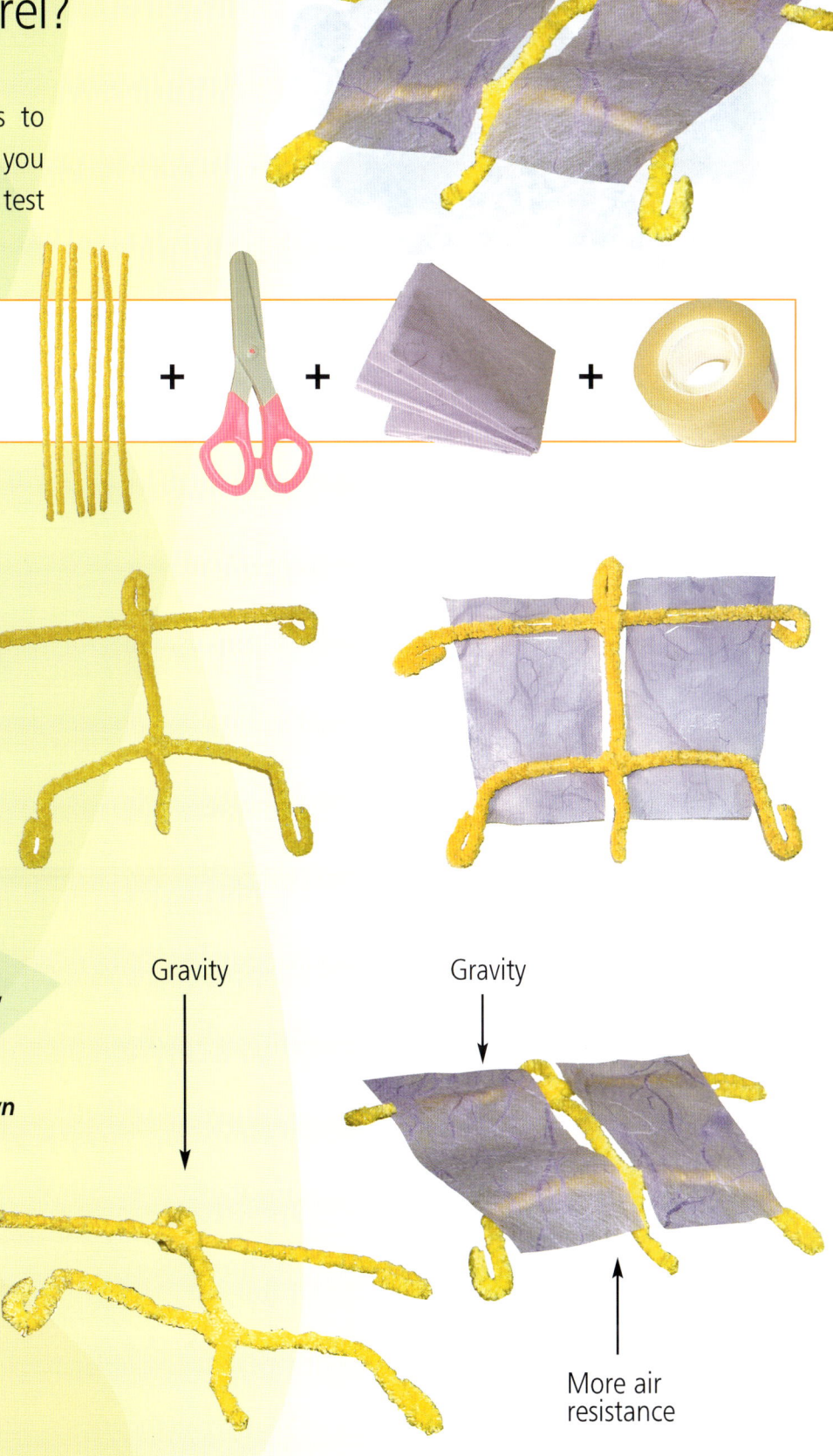

Gravity

Gravity

More air resistance

Rescue

Following a river helps a rescue aircraft navigate over the rain forest.

Before setting out, explorers tell someone where they expect to be and when they expect to return. If they don't return on time, a search may be started to rescue them. The best way to search for people in the rain forest is to fly over it and look for signs of them. Helicopters are used where there are spaces for them to hover close to the ground or land, without catching their rotors on the trees. A floatplane may be used to land on a river.

Getting noticed

Some people on an **expedition** may become ill and cannot complete their journey. They need to get into an open space where they can be seen from the air. They also need to catch the attention of a passing pilot.

Rivers make an open space in the canopy, so waiting on a riverbank increases your chance of being seen. Aircraft flying over a rain forest often follow the course of the river. In the Amazon, huge lapuna trees ("lighthouse trees") stand out from other trees on the riverbank and are used as markers by people in boats and aircraft to help them navigate. Stopping by a huge riverside tree further increases your chance of being seen by rescuers.

Why aircraft fly

When a plane moves forward, or a helicopter's rotor blades whirl around, air flows over the wings or blades. Because of the curved, **airfoil** shape of the wings or blades, the air pushes more on their underside than on the upper side. This difference in the push keeps the aircraft in the air.

Testing aircraft wing shapes

You need two strips of paper 8 in (20 cm) by 3.2 in (8 cm), two pieces of plastic straw 2.4 in (6 cm) long, two pieces of thread 10 in (25 cm) long, sticky tape.

1. Make two holes in one of the paper strips, as shown in the top photo. The holes should be large enough to push the plastic straw through.

2. Fold the paper over lightly, stick the ends together and insert a straw through the holes.

3. Rule three lines on the other paper strip and make holes, as shown.

4. Fold the paper along the lines to make a box shape, tape the ends together, and insert a straw through the holes.

5. Insert a piece of thread through the straw in each "wing."

6. In turn, hold each "wing" by the ends of the thread, about 6 in (15 cm) from your mouth, and blow. **Which one rises into the air?**

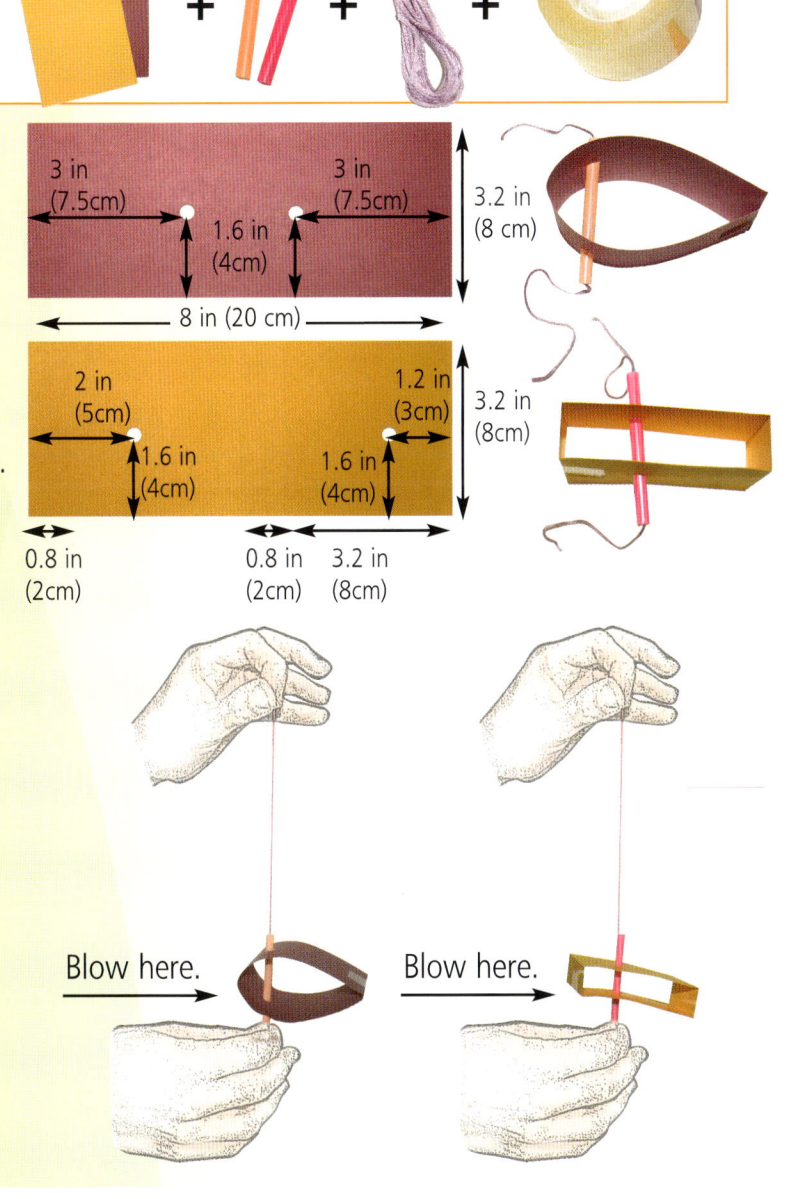

Blow here.

Blow here.

The end of the journey

At the end of a journey, people often feel that they have learned a lot along the way. How did you enjoy visiting the rain forests in this book and trying out your science skills? What would you wear in the rain forest? How would you travel and find your way? Could you tie a knot to help make a shelter? Rain forests are teeming with living things, and some of the plants and animals have survived there since the time of the dinosaurs. Maybe you will make a real journey to visit them someday.

Glossary

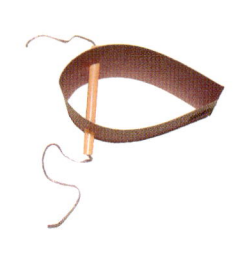

Aborigine	person who belongs to a group of people that came to live in an area of Australia thousands of years ago	**condensation**	process in which water in the form of a gas (water vapor or steam) changes into liquid water
absorb	to take in a liquid, such as water. The water may enter spaces inside the plant or material that is absorbing it.	**conduction**	movement of heat through a substance. The heat is passed from one part of the substance to another, without the substance moving. (*See also* convection currents and radiation.)
airfoil	object that has surfaces curved in such a way that, when air flows over them, the object rises	**convection**	circular flow of air or water that carries heat from a warm spot and spreads it out
bacteria	microscopic living things that feed on dead plants and animals and some materials. Some kinds of bacteria invade other living things to feed and breed and may cause disease.	**current**	part of air or water that is moving
		elasticity	ability of a material to go back to its original shape after it has been stretched or squashed
bromeliad	plant that has its leaves arranged in a ring so that they collect and hold rainwater for it to use	**evaporation**	process in which a liquid changes into a gas at normal temperatures, such as those in a room or outside
buoyancy	force that pushes upward on any object that is placed in a liquid	**expedition**	journey made for a particular purpose, such as to find out about the plants and animals in a place
buttress roots	thick roots that grow down the side of a tree trunk and form no gaps between the trunk and the ground. Often they are triangular-shaped.	**friction**	force that is produced when two surfaces rub together. The action of rubbing makes heat.

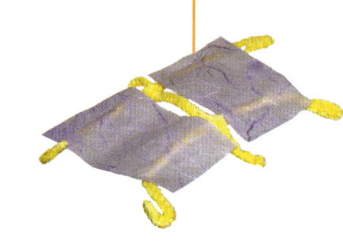

gravity	force that pulls objects down toward the center of the earth. Gravity is also a force that exists between any two objects in the universe, but is often too weak to make the objects move together.	**observation**	looking carefully at the way something is, or the way in which something happens
habitat	place where plants and animals live together	**prehensile**	able to grab hold of something for support or to pick it up
humid	word describing air that has a large amount of water vapor in it and feels damp	**property**	characteristic that a material possesses. For example, iron has the property of hardness.
improvise	to find a solution to a problem quickly, using objects and materials that are ready at hand	**prop roots**	roots with a cylindrical shape that grow out from the lower part of a tree trunk. There are gaps between the roots and the ground.
lianas	types of plants that grow in tropical regions by wrapping themselves around the trunks of trees and spreading out along their branches	**radiation**	movement of heat by waves that can pass through air and space
		ravine	deep but very narrow valley with vertical sides of rock
machete	large, heavy knife used for cutting into rain forest vegetation. In Southeast Asia it is called a parang.	**resonator**	cavity in which sounds are reflected off the inside wall. The way the sounds are reflected make them louder.
magnetism	property of a material or object that exerts forces on magnets and some metals such as iron or steel	**saliva**	liquid made in the mouths of animals to help them digest their food
		tadpole	young stage of a frog. It lives in water and has a fish-like body.
malaria	disease carried by mosquitoes. It produces fevers and can be fatal.	**tinder**	any kind of dry material, such as grass or fine wood shavings that can be used to start a fire

Index

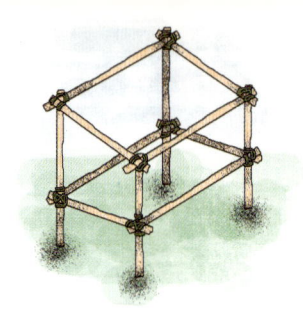

airfoil 44, 45
air 4, 10, 11, 14, 17, 18, 21, 31, 42, 43, 44
aircraft 44, 45
air resistance 43
animals 4, 6, 14, 16, 28, 30, 31, 32, 36–37, 38, 40–43
ash 30, 31

bacteria 15, 32, 38
birds 7, 16, 42
blood 15, 36, 38, 39
boat 20
boiling 30, 32, 34, 35
bridge 19
bromeliads 34
buoyancy 21

canopy 16, 19, 22, 32, 40, 42, 44
carbon 20, 31
clothes 12–15, 30, 39
compass 22, 25
condensation 10, 11, 17, 32
conduction 15, 32
convection currents 10, 32

dugout canoe 20, 21

Earth, the 5, 8, 23
equator 6, 8, 9
evaporation 9, 11, 15, 17
expeditions 4, 20, 26, 32, 34, 35, 40, 44

filtering 32, 33, 34, 35
fire 30–31, 32
fishing 20, 36
food 4, 9, 30, 36–37

forces 18, 19, 21, 27, 36, 43
friction 30
frogs 6, 7, 40, 41, 42
fruits 4, 37

gravity 18, 43

heat 8, 9, 11, 12, 14, 15, 17, 20, 30, 31
humid air 11
hunting 36, 37

insects 6, 7, 14, 28, 31, 35, 39
iron 23, 24

knots 26, 27, 29

layers of rain forest 16
leaves 9, 10, 12, 18, 28, 31, 32, 34, 35, 40
leeches 14, 39
lianas 16, 19
lice 39
light 5, 8, 9, 16, 18, 19

machete 12, 19
magnetic materials 24
magnets 22–25
malaria 31, 38
materials 13, 14, 24, 26, 28, 32
monkeys 6, 16, 19, 36, 40
mosquitoes 14, 31, 38

night 4, 17, 28

paths 7, 12, 19, 28
pitcher plants 7, 35
plant fibers 28, 29

plants 6, 9, 10, 11, 16–18, 19, 28–29, 34–35, 36, 37, 38
poison 36, 37, 38, 42

radiation 15
rain 4, 6, 9, 10, 11, 14, 28, 32, 34
rain forest people 6, 7, 12, 20, 31, 34
ravine 19
resonator 40, 41
rivers 6, 9, 11, 16, 20, 28, 32, 36, 44
roots 7, 9, 10, 18, 34

salt 15, 35, 38
shelters 26, 28
smoke 31
snakes 14, 26, 28, 38, 42
soil 10, 11, 17, 18, 32
sounds 40, 41
squirrels 7, 42, 43
sun 8, 10, 11, 12, 22
sunlight 8, 17
sweat 4, 14, 15, 35

temperature 8, 15
ticks 39
tinder 30
trees 9, 10, 11, 16, 18, 19, 20, 22, 26, 28, 34, 37, 40, 42, 44

vines 16, 19, 26, 28, 32, 34

water 4, 6, 9, 10, 11, 15, 16, 17, 18, 30, 32–35, 38
water cycle 9–11
water vapor 9, 10, 11, 15, 17, 32
wood 20, 21, 26, 30